Ecology, Culture and Health
A Primitive Tribe

Pramod Misra

SERIALS PUBLICATIONS
New Delhi

First Published–2004

ISBN: 81-86771-27-1

Published by

Head Office : **SERIALS PUBLICATIONS**
4, B.S.S. Hall, Khyber Pass Market,
Civil Lines, Delhi–110054 (India)
Phone: 23245225
Fax: 91-11-23272135
e-mail: serials@satyam.net.in

Branch Office : 4830/24, Prahlad Street, Ansari Road,
Darya Ganj, New Delhi-110002 (India)

Printed at : R. K. Offset Process, Naveen Shahdara,

PRINTED IN INDIA

Acknowledgement

I am deeply indebted to my supervisor, Prof. (Dr.) A. K. Kapoor, Professor, Department of Anthropology, University of Delhi, Delhi for his guidance and personal interest in the subject. Without his generous help my thesis would not have taken the final shape. In course of his guidance, Professor Kapoor thoroughly checked the various drafts of my thesis, patiently made corrections of language and gave many stylistic suggestions. He was more than an academic guide to me and whenever my adverse personal circumstances dejected me and prompted me to discontinue, my study, it was Professor Kapoor who stood by me and encouraged me to rise above all obstacles and keep my spirit high to complete my work successfully. He showed me the path of success and gave me courage to tread it. He applauded me at my smallest triumph and motivated me to do better. An analysis of the underlying meanings and the connecting link between the ideas was made possible by his illuminating comments. His detailed theoretical and practical knowledge helped me to develop the concepts most significant to this study, in a comparative paradigm. His useful suggestions and criticism were helpful and enhanced the conceptual and methodological clarity of this study.

I am thankful to the former Head, Prof. (Dr.) P. K. Seth, Department. of Anthropology, University of Delhi, and Prof. (Dr.) A.K. Kalla, the present Head, Department of Anthropology, University of Delhi, for their encouragement and necessary help during this study.

I am grateful to Director General, ICSSR for granting me financial aid in the form ICSSR Doctoral Fellowship for this study.

I am obliged to Dr. Satwanti Kapoor, Reader, Department. of Anthropology, University of Delhi and Dr. Gautam K. Kshatriya, Reader, Department of Anthropology, University of Delhi for giving valuable suggestions during the course of my study.

I gratefully acknowledge the help of SDO and BDO of Shahabad for granting permission to collect my data and providing me with all necessary facilities for carrying out my work, during my stay in the field.

I would like to express my sincere thanks to Mr. Neeraj Sharma, Agriculture Supervisor, Shahabad, Mr. Badan Singh Verma, Area Organiser, Adim Jati Sewak Sangh, Shahabad and Ms. Kiran Dogra, Project Officer, Lok Jumbish Pariyojna, Shahabad for their co-operation and valuable suggestions.

I am thankful to Mr. Atul Gupta for his consistent support and suggestions during the computer typing.

I would like to thank my friends Parveen Mishra, Shailesh Kumar Das, Tinu Aier and Chuba for their help and encouragement during the course of my work.

I am also thankful to my family members for giving me encouragement and moral support during the entire work. I feel highly grateful to my mother Dr. Prabha Misra and brother Vinod Misra for their loving care and affection which endured me throughout my work. They also inspired me from time to time and gave me their spiritual co-operation.

Finally, I am grateful and offer my sincere thanks to the Saharia for their co-operation without which this work would never have been completed. I am deeply thankful to all of them.

January, 2004

Dr. Pramod Misra
Department of Anthropology
University of Delhi
Delhi–110007

List of Abbreviations

ADM	-	Additional Divisional Magistrate
AIDS	-	Acquired Immuno Deficiency Syndrome
ANC	-	Antenatal Care
ANM	-	Auxiliary Nurse Mid-wife
ASFR	-	Age Specific Fertility Rate
BCG	-	Bacille Calmitte Guerine
BDO	-	Block Development Officer
BMO	-	Block Medical Officer
BPHC	-	Block Primary Health Officer
BPL	-	Below Poverty Line
CBO	-	Community Based Organization
CBR	-	Crude Birth Rate
CDPs	-	Community Development Projects
CDR	-	Crude Death Rate
CHC	-	Community Health Centre
CHW	-	Community Health Worker
CMHO	-	Chief Medical Health Officer
CSSM	-	Child Survival and Safe Motherhood
CTIs	-	Central Training Institutes
DPT	-	Deptheria Pertusis Tetanus

ENT	-	Ear, Nose and Throat
FHW	-	Female Health Worker
FP	-	Family Planning
GFR	-	General Fertility Rate
HC	-	Health Centre
HFA	-	Health For All
HIV	-	Human Immuno-defeciency Virus
HSR	-	Health Systems Research
ICDS	-	Integrated Child Development Services
ICMR	-	Indian Council of Medical Research
ICSSR	-	Indian Council of Social Science Research
ITDPs	-	Integrated Tribal Development Projects
IEC	-	Information, Education and Communication
IIPS	-	International Institute of Population Sciences
IMR	-	Infant Mortality Rate
ISM	-	Indigenous System of Medicine
IUD	-	Intra-Uterine Devices
KAP	-	Knowledge Attitude and Practices
MADA	-	Modified Area Development Approach
MCH	-	Maternal and Child Health
MFALDA	-	Marginal Farmers and Agricultural Labour Development Agencies
MHW	-	Male Health Worker
MO	-	Medical Officer
MOHFW	-	Ministry of Health and Family Welfare
MP	-	Madhya Pradesh
MPWs	-	Multi-Purpose Workers

NCAS	-	National Centre for Advocacy Studies
NFHS	-	National Family Health Survey
NFI	-	National Foundation of India
NGOs	-	Non Governmental Organizations
NHP	-	National Health Policy
NICD	-	National Institute of Communicable Diseases
NIHFW	-	National Institute of Health and Family Welfare
NIN	-	National Institute of Nutrition
NIR	-	National Institute of Rural Development
OADR	-	Old Age Dependency Ratio
OBC	-	Other Backward Classes
OPV	-	Oral Polio Vaccination
ORT	-	Oral Rehydration Therapy
OTCS	-	Orientation Training Centres
PHC	-	Primary Health Centre
PMP	-	Private Medical Practitioner
PNC	-	Postnatal Care
PTGs	-	Primitive Tribal Groups
RCH	-	Reproductive and Child Health
RMRC	-	Regional Medical Research Centre
RTIs	-	Respiratory Tract Infections
SC	-	Sub-Centre
SDH	-	Sub-Divisional Hospital
SDO	-	Sub-Divisional Officer
SFDA	-	Small Farmers Development Agencies
ST	-	Scheduled Tribe
SDP	-	Saharia Development Project

SRS	-	Sample Registration System
STDs	-	Sexually Transmitted Diseases
TBA	-	Traditional Birth Attendant
TDR	-	Total Dependency Ratio
TFR	-	Total Fertility Rate
TSP	-	Tribal Sub-Plan
UCI	-	Universal Child Immunization
UIP	-	Universal Immunization Programme
UN	-	United Nations
UNICEF	-	United Nations International Children's Emergency Fund
UP	-	Uttar Pradesh
UTS	-	Union Territories
VANI	-	Voluntary Action Network of India
VHAI	-	Voluntary Health Association of India
VHW	-	Village Health Worker
WHO	-	World Health Organization
WSSD	-	World Summit on Social Development
YADR	-	Young Age Dependency Ratio

Contents

Acknowledgement *v*

List of Abbreviations *ix*

1. Introduction 1
2. Area and People 51
3. Methodology 66
4. Socio-Cultural Life of Saharia 100
5. Health Status and Practices 135
6. Summary, Conclusion and Suggestions 267

References 317

1

Introduction

Much of the development of medical anthropology has occurred since World War II. The study of human confrontations with disease and illness, and of the adaptive arrangements (that is, medicines and medical systems) made by human groups for dealing with these ever-present and panhuman dangers, has come to be known as medical anthropology (Landy, 1977). By the early 1960s the term medical anthropology had come to be used increasingly by anthropologists working in and around problems in health and disease in human societies. Although the term medical anthropology was not in general use before Scotch (1963). The term was used by many scholars (Caudill, 1953), (Polgar, 1962), (Scotch, 1963), (Hughes, 1968), (Fabrega, 1972), (Lieban, 1973), (Foster and Anderson, 1974).

Medical anthropology encompasses the study of medical phenomenon as they are influenced by social and cultural features, and social and cultural phenomena as they are illuminated by their medical aspects (Lieban, 1974).

Medical anthropology includes interests and contributions from all the branches of anthropology.

Without this broad, bio-cultural approach, the use of the ecological framework, already so promising in other sub-disciplines of anthropology and which has made an initial impact in the medical anthropological area, will not be utilized.

Compared to most other disciplines, the hallmark of anthropology, medical anthropology included, is the so-called holistic approach. The medical anthropologists collect a great deal of data about economic features, social relationships, cultural belief systems, political processes, and other aspects of a community, even if the research intention is focused on a specific health question.

Medical anthropology deals with people's point of view in medicine. There are different schools of thought in medical anthropology. These are culture trait approach (Clement, 1932); functional approach (Ackernecht, 1958; Mead, 1953, Foster, 1979); Paul, 1955); symbolic approach (Turner, 1964); semantic approach (Good, 1977; Fabrega, 1978; Kleinmann and Sung, 1979).

Medical anthropology is concerned with the biocultural understanding of man and his works in relation to health and medicine (Alland, 1966; Roney, 1959, 1963; Scotch, 1963). While this biocultural orientation is basic to the whole field of anthropology, medical anthropology as a subspecialty or as a "synthetic discipline' with multidisciplinary orientations may involve medical scientists, anthropologists, biologists, and sociologists (Roney, 1959,1963).

It has been observed that medical anthropology is the holistic study of health, illness and related misfortunes, as these are culturally perceived, labelled, classified, experienced and communicated on one hand and socially constructed roles, statuses and institutional networks which are believed to help in the health enhancing process, on the other, with a view to identify cross-cultural similarities and variations in the patterning of such behaviour.

Medical anthropology, therefore, is a bio-cultural discipline concerned with both biological and socio-cultural aspects of human behaviour, particularly with the ways in which the two interact and have interacted throughout human history to influence the conceptions of health and treatments and therapies of disease (Foster and Anderson, 1978:2-3).

The Ecological Interests of Medical Anthropologist's

Medical anthropologists are ecologically oriented and are concerned with the interrelationships between man's natural and social environments, his behaviour, his diseases, and the ways in which his behaviour and diseases have influenced his evolution and culture through feedback processes. Environment is both natural and socio-cultural. All groups must adjust to the geographical and climatic conditions that prevail in the places they live, and they must learn to exploit the resources available to meet their needs. Disease, for example, is a part of the human environment. Diseases involve pathology and, at one level, it is obviously biological. Yet socio-psychological and cultural factors often play roles in triggering disease, while the way in which a patient's environment is altered while he is undergoing care is purely cultural (Foster and Anderson, 1978).

CULTURE

Culture is that complex whole which includes knowledge, belief, arts, morals, law, custom and any other capabilities and habits acquired by man as a member of society (Tylor, 1871).

Culture is all those means whose forms are not under direct genetic control which serve to adjust individuals and groups within their ecological communities (Binford, 1968).

Culture consists of the shared practices and understandings within a society. To some degree culture is based on shared meanings that are to some extent "public" and thus beyond the mind of any individual (Geertz, 1973).

Malinowski (1922) mentioned that every living culture is a functional and integrated whole analogous to an organism. No part of a culture may be understood except in relation to the whole.

Culture is the man-made part of environment (Herskovits, 1952). Two of the most basic components of culture are material culture and non-material culture.

Material culture consists of the physical products of human society (ranging from weapons to clothing styles, whereas non-material culture refers to the intangible products of human society (values, beliefs and norms).

Cultural Explanation of Ill Health

Disease did not begin with man, but in a sense as Sigerist noted, disease is a part of life itself; that is, a manifestation of life under altered conditions. Cultures are the responses made by human groups to obstacles placed by sickness and trauma in the path of successful adaptation. Theories of disease (scientific or religious) including etiology, diagnosis, prognosis, treatment and amelioration or cure, all are part of the cultural repertory and equipment of human groups. Culture plays a major role in determining the patterns of disease and death in a population, for two reasons. First, culture may shape important behaviours (with respect to diet, activity patterns, sexual practices and so forth) that predispose the individuals to certain diseases. Second, through culture, people change the nature of their environment, often in ways that affect their health.

Every culture has its particular explanations for ill health. Cultures provide people with ways of thinking that are "simultaneously models of and models for reality". Geertz (1973) and Frake (1961) in the diagnosis of disease among Subanum of Mindanao, described sickness as a vehicle for pursuing other interests. Like Frake, Good (1977) followed the explanatory model approach. He wrote "illness is fundamentally semantic or meaningful and all... clinical practice is inherently interpretive......". But there are fundamental differences between Frake and Good regarding how the language of sickness ought to be interpreted. According to Good, semantic illness network is the network of word, situations, symptoms and feelings which are associated with an illness and give it meaning for the sufferer (Good,1977:40). In his analysis of Iranian Semantic Illness networks, Good introduced the notion that networks are organised through core symbolical elements. Good's idea of core symbols closely parallels Turner's notion that there are dominant symbols which organise the meanings of rituals in pre-industrial societies. Semantic illness networks are also described in the work of Blumenhagen (1980:198) and Kleinman (1978). Blumenhagen (1980) describes how hypertension is perceived by a population of Americans who suffer from this disease. Blumenhagen organised his informants's responses into what he calls "nodes" and "arrows". Nodes consist of reports of symptoms, interaction, physiological functions, body states, pathogenic agencies and activities. Arrows identify casual relations between nodes. Good's work is different from earlier, more traditional anthropological views of sickness. He thinks that an informant's statements need to be interpreted in the context of his illness experience and close attention needs to be interpreted to the way his statements change over time in response to his circumstances. According to Good (1977), semantic illness networks are inseparable from the idea that illness is an individualized process. In Kleinman's work, semantic illness networks have been made part of a comprehensive framework (1978).

Sickness is a blanket term to label events involving disease and/or illness. Sickness is the domain of medical anthropologists, though their special contribution will be mainly with regard to illness (Kleinman, 1980:71-72). Semantic networks are related to system of medical knowledge through people's explanatory models of illness (E.M.). An 'EM' is a set of beliefs which contain any or all of five issues: etiology, onset of symptoms, pathophysiology, course of sickness (severity and type of sick role), and treatment (Kleinman, 1978: 87-88). The EM concept explains that every culture has its particular explanations for sickness. It simultaneously creates order and meaning, give plan for purposive action and helps to produce the conditions required for their own perpetuation or revision. Kleinman attributes EMs to individuals, and not to cultures. (Kleinman, 1980:108).

While Kleinman and his associates have been establishing the EM approach, other anthropologists found that social relations which primarily produced the forms and distribution of sickness in society.

"Disease-illness" scheme by Frankenberg and Young (1955) explains that disease retains its original meaning (organic pathologies and abnormalities) and illness refers to how disease and sickness are brought into the individual consciousness. However, sickness is no longer blanket term referring to disease and/or illness. Sickness is redefined as the process through which worrisome behavioural and biological signs, particularly ones originating in disease, are given socially recognizable meanings, i.e. they are made into symptoms and socially significant outcomes. Every culture has rules for translating signs into symptoms for linking symptomatologies to etiologies and interventions, and for using the evidence provided by intervention to confirm translations and legitimize outcomes (Comaroff, 1982; Lock, 1982; Nichter, 1980). Sickness is a process for socializing disease and illness.

1. In "pluralistic" medical systems, a single set of signs can designate more than one sickness and social forces help to determine which people get which sicknesses.
2. In western society, the right to translate signs into socially important symptoms is dominated by a single kind of practitioner and a more or less unified set of etiologies.
3. Symbols of healing are simultaneously symbols of power. Specific views of social order are embedded in medical beliefs, where they are often encoded in etiologies and beliefs about the sources of healing power.

Different Stages of Illness

As already explained, generally, any illness may have five processes i.e. (1) onset; (2) recognition; (3) diagnosis; (4) treatment and (5) outcome. In most of the cases, only at the diagnosis level, the illness may be made public, that too only in some cases some sort of health care may be sought; only those who approach formal systems of health are (mostly Government run allopathic facilities) recorded. To generalise the health situation of the population and to understand these intricacies it is essential to examine the various types of health care employed at different stages of illness.

Types of Health Care

Broadly health care may be divided into three categories i.e. (a) Informal Health Care; (b) Interface Health Care; and (c) Formal Health Care.

(a) **Informal Health Care**: Informal Health Care is provided by non-professionals such as family

members for whom these activities do not constitute an occupation, at least not a paid one. At the functional level, illness provides an occasion for reinforcing solidarity amongst relatives, friends and all others concerned, treatment is medicated through cultural mechanisms which interact with biological factors. In this form of health care, the important elements are self care, family, and the community.

Self care is one major aspect that involves self diagnosis and self treatment in the face of perceived symptoms. Self treatment can take many forms, occasionally bizarre, but more frequently mundane such as rest, dietary change (local customary foods) or the use of drugs, both previously prescribed and over the counter. Another form of self care, not usually identified as such, is doing nothing, just letting nature take its course without interference, often the best form of 'treatment" for self limiting conditions (Haug etal, 1989). Another type of informal care is that provided by family members, which may indeed be intertwined with self care for sick adults.

(b) **Interface Health Care:** Some occupations are on the border line. These are traditional healers, pharmacists in drug stores etc. They work for pay, but are not scientifically trained, although they may have a wealth of experience and shrewd perceptions of what treatment will work.

(c) **Formal Health Care:** Formal Health Care can be provided by a variety of professionals (from various systems of medicine) – nurses, physical therapists and physicians etc.

Therefore, sickness is perceived in biological, social, cultural and psychological terms in a holistic frame work.

HEALTH

India is a signatory to the Alma Ata World Health Organisation (WHO) declaration of 1978 and committed to attaining the goal of "Health for All" by 2000 A.D. through primary health care approach. The World Health Organisation's constitution (WHO 1946) defines health as 'a state of complete physical, mental and social well-being and not merely the absence of disease and infirmity'. What would a person in a state of complete physical, mental and social well-being be like? Physical well-being means that they would be free of all physical disease and free of aches, pains and other symptoms. Their joints, muscles, heart, lungs, liver, stomach and all their other organs would function well. They would be fit, strong and agile, with good eyesight and acute hearing. They would be able to resist infection and if they suffered injury they would recover quickly. Furthermore, there would be no indication that they were likely to develop any disease or infirmity in the future.

Mental well-being means that their mental function would be as good as their physical state. They would be intelligent, well-balanced, resilient and able to cope with the stresses of daily life. They would feel happy and at peace with the world and themselves.

Social well-being means that they would function well in society. They would relate well to their family and other members of society. They would have a role in society and would value and be valued by others.

This paragon of health would live in a situation which enhanced and protected their health. The physical environment would supply shelter, warmth and food. The society in which they lived would give them affection and esteem but also enough challenge to keep them stimulated.

This description suggests that 'complete physical, mental and social well-being' is not a realistic goal for health promotion. Three important points can however be drawn from the WHO definition:

1. There is a lot more to being healthy than just being in good physical shape. We also have to think of mental and social function.

2. Absence of disease and infirmity is a necessary feature of health.

3. Health requires the presence of various features which are referred to as positive health.

Table 1 shows some of the components of Health. A more practical definition of health recognizes that people differ and have different limitations. Health is a state in which people are able to work towards attaining their 'realistic chosen and biologic potentials'.

The WHO definition of health distinguishes a positive and a negative dimension to health. Negative health, or ill-health, has a subjective component expressed in the concepts of illness and discomfort, and an objective component expressed in concepts of disease, injury, handicap, or deformity. These strands are linked via the idea of abnormal, unwanted, or incapacitating states of a biological system, which in turn presupposes the idea of a good or flourishing human life. These ideas can be carried over to the spheres of the mental and social. Positive health has two components: well-being and fitness. Well-being can be purely subjective, but to be a component of positive health it must arise from and reflect a process of empowerment, which may be enhanced through the development of lifeskills.

The two concepts or similar ideas have generally taken the form of a continuum (Catford, 1983) with extremes of well-being and ill-health at the positive and negative ends, respectively (Fig.1)

Table 1: Some of the Components of Health

	Absence of Disease	*Absence of Ill Health*	*Presence of Positive Health (well-being)*
Physical Health	Long life No physical disease Low-risk of disease Low-risk lifestyle	No physical symptoms No physical disability	Fitness Health-promoting lifestyle
Mental Health	No psychiatric disease	No mental distress No mental disability	Self-esteem Mental ability Resilience
Social Health	No social/family breakdown	No social/family frictions	Role in society Social support Sense of belonging

Fig. 1: Health As a Continuum

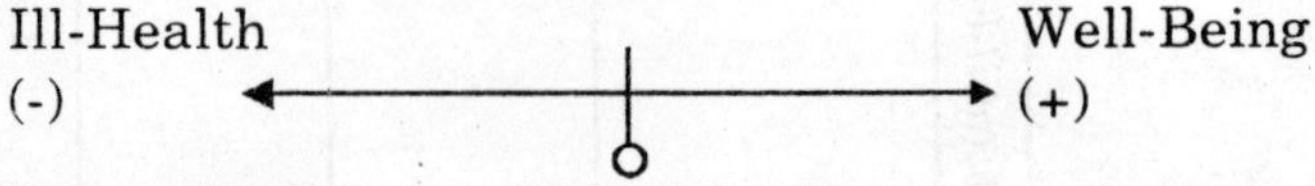

A Model of Health

The model shown as Fig.2 is an attempt to summarize the elements of health and their interrelationships.

Fig. 2 A Model of Health

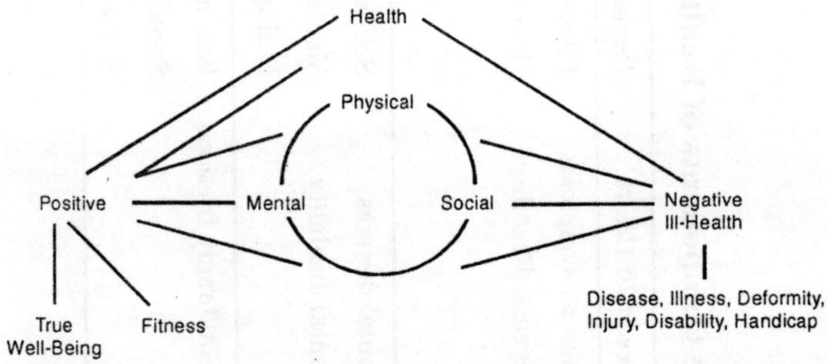

Both positive and negative health is to be seen as having interconnected physical, mental, and social elements. The representation of these facets as bridges between positive health and ill-health in relation to one particular facet often go hand in hand; physical illness may well be accompanied by a low level of physical well-being or fitness.

There are three main factors which determine how healthy a person will be (Smith, 1992):

- genetics
- environment
- lifestyle

It is helpful to think of these three things separately although they all interact and are not independent of each other. Genetic endowment cannot be changed but how the individual is affected by their genes will to a large degree be influenced by the way they live their life. It is convenient to divide living conditions into environment and lifestyle. Environment refers to aspects of a person's surroundings such as their housing, their neighbourhood, the water they drink and the air they breathe. In general these are things over which individuals have little control. Lifestyle refers to aspects of the way people live over which they have some control, such as smoking, eating and drinking. The distinction between environment and lifestyle is not clear cut and in many situations people have little real choice over lifestyle issues. The environment has a major influence on lifestyle and the lifestyle can modify the effect of the environment. Some of the major influences on health are listed in Table 2.

Ecological Approach to Health

The concept of "environment" is complex and all-embracing; it is not only physical factors like air, water and soil that form our environment but also the social and economic conditions within which we live. The environment is defined as "the aggregate of all external conditions and influences affecting the life and development of an organism, human behaviour or society" (Leavell and Clark, 1965). The internal and external environments comprise the total environment.

According to the ecological approach, disease represents a maladjustment of the human organism to his environment which is getting more complicated as man is becoming more ingenious. Human behaviour and their habits also constitute environment we are talking about. It has been established that "success of public health programmes often depends upon modifications in human behaviour".

Table 2: Some of the Major Influences on Health

Genetic make up	
Basic needs	Warmth and Shelter Food and Water Freedom from Physical Violence.
Physical environment	Housing Drinking Water/ Sewerage Air Quality Food Quality and Safety Roads Transport Noise Workplace
Lifestyle	Smoking Eating Drinking Exercise Illicit Drug Use Safe Sex Practices Use of Medical Services
Environmental constraints on lifestyle	Available of Facilities (exercise, food choice in shops) Peer Pressure Advertising Cultural Pressures Education

Health and disease are measures of the effectiveness with which human groups, combining cultural and biological resources, adapt to their environments (Lieban, 1973).

Health is never static; it is dynamic and "fluctuates within a range". What is maximum today may be minimum tomorrow. The state of health varies from day to day. In this context, health is defined as "a flexible state of body and mind which may be described in terms of a range, within which a person may sway from the condition wherein he is at the peak of enjoyment of physical, mental and emotional experiences, having regard to environment, age, sex and other biological characteristics due to the operation of internal or external stimuli and can regain that position without outside aid" (Seal, 1963). Park (1972) defines health as "a state of relative equilibrium of body form and function which results from its successful dynamic adjustment to forces tending to disturb it. It is not passive interplay between body substance and forces impinging upon it but an active response of body working towards adjustment". These definitions envisage health as a pendulum oscillating between a range or spectrum, one end of which represents the minimum and the other end the maximum. It implies that health is a `state' not to be attained once and for all but ever to be renewed. There are degrees of well-being as there are degrees of severity of illness (Figure 3).

Fig. 3: The Health-Sickness Spectrum

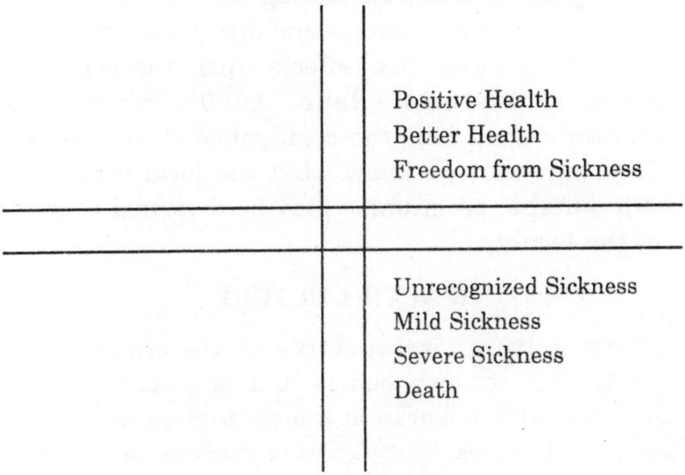

Ecology is a key-word in present-day health philosophy. Ecology is defined as the science of the mutual relationship between living organisms and their environment. "Human ecology is concerned with the broad setting of man in his environment" (Hanlon, 1969). Ecology of health is the study of the relationship between variations in man's environment and his state of health. The basic theme of ecology "is that everything is related to everything else".

According to the ecological approach, health is a state of dynamic equilibrium or adjustment between man and his environment. One can think of this graphically as a balanced scale with the pans representing the agent and human host and the flucrum, the environment and health as a state of equilibrium between the disease agent and the human host. When this balance is disturbed for any reason, ill-health results (Leavell and Clark, 1965).

Thus the trend in recent times has shifted from the study of 'disease in man' to 'Man in the midst of Disease'. In other words it is the study of man in his environment. As new ways of living have been developed by man, new ecological systems, including pathogens, have been formed. Urbanization, industrialization and other patterns of social organization all have their effects upon the composition of ecological systems (Rene, 1969). Since health programmes should take the cognigance of local practices it is important that we know what the local practices are so that a suitable programme may be designed to suit the need of the locality.

HEALTH CULTURE

Every culture, irrespective of its simplicity and complexity, has its own beliefs and practices concerning diseases. No culture works in a meaningless fashion in its treatment of diseases. Every culture evolves its own system of medicine in order to treat diseases in its own way. This treatment of disease varies from group to group. In every society, there exists a body of beliefs and concept with regard to the nature of disease and its treatment. One common feature, however, is there close integration with the other institutions and organizations of society. Cultural patterns and religious beliefs, economy and morality, social values and medical beliefs are all found together to form the 'health-culture' of a community. The concept of health culture refers to a sub culture within a totality of population. It encompasses a vast complex of knowledge, beliefs, techniques, roles, norms, values, ideologies, attitudes, customs, rituals and symbols which are related to health and disease. That means the socio-cultural aspects of health, the beliefs and practices relating to health and disease, the diagnosis and treatment methods, the healers and curers and their recruitment, concepts and organization of medical system etc. In simple terms the subject of health culture focuses on the nature of illness as it is conceived by the natives, their own methods and criteria of classification of disease, the causes and cures, types of therapists that seek to alleviate illness and their skills and social roles, preventive measures, the relation between magic and religion, cultural aspects of ethnomedicine, introduction of western medicine into traditional villages, illness behaviour and finally the ethnopsychiatry. As Banerji (1982) says the culture of a community determines its health culture–that is, cultural meaning of the health problems of the community, and the means the community adopts to deal with them.

This health culture forms one of the two dimensions of medical anthropology. The first dimension is anthropology in medicine which emphasizes the contribution of anthropological knowledge to the diagnosis and treatment of disease, the second is anthropology of medicine, which concerns itself with anthropological study of the medical profession. Hence the study of health culture contributes to the understanding of contributions of anthropological knowledge of the etiology, theory of diseases, diagnosis, treatment of diseases and the place medicine occupies in the lives of the people.

Health problems and health practices of any community are profoundly influenced by interplay of a complex of social, economic and political factors. To analyse the psycho-anthropological aspects of health culture it is necessary to take a holistic view of all the cultural dimensions that are related to the overall culture of the community. For developing such a view it is necessary to consider simultaneously the various cultural factors which generate certain health problems in a community, the cultural perception and cultural meaning of various institutions that the community has either a diffusion or an intrinsic innovation (eg. Shamanism, withchcraft, spiritual healing etc.) by the community (Sahu, 1986).

TRIBES IN INDIA

According to article 342 of the Constitution, the Scheduled Tribes are the tribes or tribal communities which may be notified by the President. The census enumerates only such tribal committees as are scheduled under the relevant constitutional order in force at the time of the Census.

India has the largest tribal population in the World. As per the 1991 census the tribal population of India was over 67.8 million constituting about 8.1% of the country's population. This figure of Scheduled Tribes excluded Jammu and Kashmir where the 1991 census enumeration did not take place. Table 3 shows the distribution of scheduled tribe population in different states/UT and their percentage with respect to (a) total tribal population in India (b) total population of their respective states/UTs and

Table 3: Distribution of Scheduled Tribe Population over States/Uts

State/UT	*Total Population*	*ST Population*	*% share to ST Population*	*% ST to State/UT Population*	*No. of STs*
1	2	3	4	5	6
India	846,302,688	67,758,380	8.01	–	573
A.P.	66,508,008	4,199,481	6.47	6.31	33
Arunachal	864,558	550,351	0.85	63.66	12
Assam	22,414,322	2,874,441	N.A.	N.A.	14
Bihar	86,374,465	6,616,914	10.20	7.66	30
Goa	1,169,793	376	Neg.	0.03	5
Gujarat	41,309,582	6,161,775	9.50	4.22	29
Haryana	16,463,648	–	–	–	–
Himachal	5,170,877	218,439	0.34	4.22	8
J&K	7,718,700	N.A.	N.A.	N.A.	8
Karnataka	44,977,021	1,915,691	2.95	4.26	49
Kerala	29,098.518	320,967	0.47	1.10	35
M.P.	60,181.170	15,399,034	23.73	23.27	46
Maharashtra	78,937,187	7,218,281	11.28	9.27	47
Manipur	1,837,149	362,173	0.97	34.41	29
Meghalaya	1,774,778	1,517,927	3.34	85.53	17

(Table Contd...)

(Table 3: Contd...)

1	2	3	4	5	6
Mizoram	689,756	653,565	1.01	94.75	14
Nagaland	1,209,546	1,060,822	1.64	87.70	5
Orissa	31,659,736	7,032,214	10.84	22.21	62
Punjab	20,281,969	–	–	–	–
Rajasthan	44,005,990	5,474,881	8.44	12.44	12
Sikkim	406,457	90,901	0.14	22.35	2
Tamilnadu	55,858,946	574,194	0.88	1.03	36
Tripura	2,757,205	853,345	1.32	30.95	19
Uttar Pradesh	139,112,287	287,901	0.44	0.21	5
West Bengal	68,077,965	3,808,760	5.87	5.60	38
Andaman & Nicobar	280,661	26,770	0.04	9.54	6
Chandigarh	642,015	–	–	–	–
Dadra & Nagar Haveli	138,477	109,380	0.17	78.99	7
Daman & Diu	101,586	11,724	0.02	11.54	5
Delhi	9,420,744	–	–	–	–
Lakshadwip	51,747	48,163	0.07	93.15	–
Pondichery	807,785	–	–	–	–

Source: Census of India (1991)

(c) number of varieties/types of STs recognized in the state/ UTs. It may be noticed from there that the population of scheduled tribes varies a great deal from one state to another. Their largest population is found in undivided Madhya Pradesh (15,399,034). The second largest number of the tribals is enumerated in Orissa (7,032,214) immediately followed by undivided Bihar (6,616,914). But their largest proportion of total population among all the states is found in Mizoram (94.75%) followed by Nagland (87.70%), Meghalaya (85.3%) and Arunachal Pradesh (63.66%). Among Union Territory Lakshadweep (93.15%) ranks first followed by Dadra Nagar Haveili (78.99%). The tribals in the 1991 census were enumerated throughout the country except in two states viz., Haryana & Punjab and three UTs viz., Delhi, Chandigarh & Pondichery where no Scheduled Tribes have been notified by the President of India. As on today, the estimated/project tribal population is over 80 million. The tribal communities belong to different ethnolingual groups, profess diverse faith and are at various levels of socio-economic development. Spread along the entire spectrum ranging from hunters and gatherers of forest produce to the urbanised skilled or industrial wage earners—tribal communities constitute very important segment of the Indian population. The size of the tribal communities varies greatly. The tribal population of India belong to three racial groups-Negritoes, Austroloids and Mongoloids. Linguistically, the tribals are classified into four major language families-Austro Asiatic, Tibeto Chinese, Indo European and Dravidian. The tribals in India, speak over 105 different languages with 225 subsidiary ones. The spatial distribution of the tribal communities is characterized by the striking tendencies to cluster in a few pockets of diverse degree of isolation. Ecologically, the tribal homelands are far from homogenous and as such display a diversity of high order. The areas of tribal concentration have been generally described as the forest and hilly areas of the country. No systematic classification of the tribal areas had been attempted untill beginning of the 5th plan period when a new strategy for tribal development was evolved and the areas of tribal concentration were systematically identified and demarcated on some objective criteria. A clear picture of the areas of the tribal concentration in the national scene has emerged only thereafter. Now the tribal areas of India can be broadly divided into 6 regions viz., South Central, North Central, North Western, North Eastern, Southern and Oceanic groups. Each region has some distinguishing characteristics of its own. They differ considerably amongst themselves in terms of geo-climatic condition, resource potential and demographic characteristics (Ali, 1992; 2001).

PRIMITIVE TRIBAL GROUPS (PTGs)

The primitive tribal groups are characterised by pre-agricultural level of technology; low level of literacy and stagnant or diminishing population. There are 75 primitive tribal groups in India living in different ecological zones.

Singh (2000) and Patnaik (1993) have discussed about the Primitive Tribal Groups (PTGs).

According to Singh (2000) primitive tribal groups are integral parts of nature. Birth, life and death formed a continuum. The living and the dead were together in a wholesome relationship. Ancestors did not vanish from a community's world when they died; they acquired a niche in the supernatural world with which the living were in direct communion. Thus the supernatural, nature and life where movements of a grand symphony evocative of rhythm and harmony. The banal and the sublime pulsated in unison. Stridency and discordance were at a discount.

In recent years the concept 'primitive' has been subjected to increasing criticism from a number of anthropological quarters. The term 'primitive' is considered a cliché and a damaging word. It is not only confusing and damaging but also unreal and obfuscating. Many communities are labelled primitive irrespective of manifold varieties which they present in their social organisations and cultural pattern. As it is unjustified to call all kinds of trees in a forest as Oak trees, likewise it is a mistake to designate different communities as primitive. In the case of the former all those trees we call Oak have no characteristics of Oak tree in common among them and in the case of the latter all the different communities we call primitive have no social or culture qualities in common to be designated by a common term (Patnaik, 1993).

According to Patnaik (1993) "when we think of term primitive it means that whatever has preceded us is less advanced. Of course it is true in many respects. For example, it is true of trading and writing and of technological progress. But in many other respects the cultures of those whom we call primitive is more highly developed than are the civilized or advanced cultures. For instance, the Juang of Keonjhar district, the Saora of Ganjam Agency and the Bonda highlanders of Koraput district are called primitive tribal communities. But their cultural patters and social structures, economic and ritual life and their cognitive orientations reveal that they are more generous, hospitable, loving, cooperative, honest, dependable, cheerful and courageous than most of the members of advanced or civilized societies". The term primitive denotes a variety of terms which may be arranged under seven major aspects (See Table 4).

It appears from the above statement that most of the expressions pertaining to the term primitive do not seem to have any significant intellectual or analytic advantage. Therefore the concept primitive may have neither empirical validity nor theoretical utility and our use of this term is through mere force of habit and has been intellectually

Table 4: The Term Primitive Denotes Seven Major Aspects

	Aspects		*Connotations of the term Primitive*
1.	Communication	-	Isolated, arrested in development, folk society.
2.	Economics	-	Moneyless, lack of economic specialisation, cooperation for common goals.
3.	Education	-	Non-literate, lacking historical records, lacking literature, lack of abstract time reckoning.
4.	Personality	-	Endowed with over-powering sense of reality, everyday facts associated with religious and ritual performances, uniformity of language and religion and strong concept of personal security.
5.	Size	-	Small-scale
6.	Social	-	Relationship based on kinship, relative homogeneous, strong political and social organizations.
7.	Technology	-	Simple, simple tools, not civilized, pre-historic, aboriginal, non-industrialized, savage, non-urban, tribal and traditional.

unproductive. It is also a sheer mistake to think of primitive as opposed to civilized or advanced. There is no empirically defensible ground for dichotomizing the culture or societies into such paired opposites as illustrated above. Some small societies which are called primitive are endowed with highly sophisticated urban qualities. The kinship system of a highly complex industrialized society may be extremely simple but kinship system of a tribal community labelled as primitive may be the most complex in the world. In short, the primitive society is a community, springing from common origin, composed of reciprocating persons and growing from within. It is not a collective; collectives emerge in advanced societies; they are functional to specialized ends and they generate a sense of being imposed from outside. These prominent features of primitive society may be ideal but they will surely help us in identifying contemporary tribal communities who approximate to these ideal primitive characters and which deviate from such characters. And it is on this basis largely can a primitive-civilized continuum be reconstructed and the positions which the different tribal communities will occupy in this continuum determined.

According to Singh (2000) "there is a need to distinguish a category of tribal communities who have been regarded as 'primitive' in the classical sense of the term. Singh (2000) has classified Primitive Tribal Groups (PTGs) into Primary Primitives, Marginal, SubMarginal and Subsistence Groups".

Primary Primitive Groups

Primary primitive groups have been happily placed in an environment more or less insulated from rough and aggressive external forces. Even if they have traded or made cultural contact with other groups and communities, this contact has been minimal. Examples of such groups are the Jarawa and Sentinelese of the Andaman and Nicobar Islands, the Bondo, Lanjia Saora and Kutia Kondh of Orissa, the Cholanaicken of Kerala and the Abhuj-Madia of Madhya Pradesh. On the other hand, since the marginal tribal communities live on the fringes of the larger society, some interaction is inevitable. It is even conceivable that their present state has been a consequence of such prolonged interaction. Some social scientists have described the interaction as regular economic symbiosis with the market economy of the larger society.

Subsistence Marginal Groups

Some, shifting cultivators live tenuously on farm and forest produce which needs supplementation. A number of tribes fall in this subsistence category, like the Hill-Kharia of Orissa, the Gadava of Orissa, the Chenchu of

Andhra Pradesh, and the Baiga of Madhya Pradesh. The extent of their interaction and confrontation with the neighbouring population has varied, but it has not been so significant as to throw them overboard in search of other avocations. This category may be called subsistence marginal groups.

Sub-Marginal Groups

Other shifting cultivators like the Birjia of Palamau district of Bihar and the Bharia of Patalkot in Madhya Pradesh have had to abandon even shifting cultivation more and more to depend on minor forest produce and landless labour. The ecological base of the habitat of some others has been seriously eroded. Forests have mostly disappeared and so has land of good quality. The Paharia of Santhal Parganas, Bihar and the Birhor of northern Orissa and southern Bihar exemplify this. They live by hunting and gathering and as landless labourers. They have preserved their identity but not well-defined locality. They may be regarded as sub-marginal groups.

Subsistence Groups

A slot beyond, the Hill Kharia of Purulia district of West Bengal seem to have almost lost control over forest resources, unlike the Birhor of Orissa and the Cholanaicken of the Nilambur Valley of Kerala. Their minimal reliance on the forest and predominant dependence on non-forest occupations does not place them in geographic marginality to the adjoining population, though economically they may subsist on the margin. Hence, they may be regarded as belonging to a subsistency category (c.f. Singh, 2000).

TRIBAL HEALTH STATUS: AN OVERVIEW

It has been observed that poor health status reflected in tribal health care practices is the outcome of their ecological, social and living conditions, illiteracy, ignorance, under availability of potable water etc. and in appropriate medical and health opportunities. Indigenous and modern health care system of practices continues to be antagonistic to each other. A brief review on the available health indicators on various tribal population groups and their implications in health care is provided. There is a general agreement that health status of the tribal population in India is very poor. Different studies (Basu, 1986, 1990; Chowdhuri et al, 1986; Mukherjee, 1986, 1990; Rizvi, 1986; Sen et al, 1986; Haque, 1990; Mahapatra et al, 1990; Swain et al, 1990) have tried to establish this with the help of morbidity, mortality and health statistics. The widespread poverty, illiteracy, malnutrition, absence of safe drinking water and sanitary and living conditions, poor maternal and child health services, ineffective coverage of national health and nutritional services, have been traced out in several studies as possible contributing factors for dismal health conditions prevailing among these vulnerable populations (Bhasin, 1990). It has been found that certain diseases like goiter, yaws, malaria and guinea worm are endemic in tribal areas like Madhya Pradesh, Orissa, Rajasthan, and Gujarat (Working Group Report of Ministry of welfare for ST, 1998). Besides these, genetic disorders have been found to occur in high frequency among various tribal groups and scheduled caste populations. This situation is further complicated by the prevalence of high frequency of consanguinity found to be practiced by different tribal population e.g. Paniyan tribal group of Kerala shows 32 per cent of consanguinity (Basu, 1993).

Demographic Trends

Crude Birth Rate shows considerable heterogeneity among different tribal groups. Crude Birth Rate varies from a minimum of 31.2 in Lodhas (Basu,1999) of West

Bengal to a maximum of 75 in Soliga of Karnataka (Murthy,1987). Crude Birth Rate is invariably above 30 in most of the tribal groups in India.

Crude Death Rate varies from a minimum of 9 in Hill Korwas (Pandey,1993) to a maximum of 25 in Oraons of Bihar (Parsuram etal,1990). Data reflect a large range in Crude Death Rate. This could be due to investigations being carried out in different years and in different geo-climatic tribal habitat and also due to inadequate systematic, well designed study. Nevertheless, a very low rate of rural tribal female literacy could be one of the major attributable factors for a high Crude Birth Rate and Crude Death Rate in all the tribal populations besides other social, cultural, economical and technological factors.

Fertility rates gave a variable picture among the Indian tribes. Total Fertility Rate varies from 3.14 among Mirdhas of Orissa (Chaudhuri etal,1994) to 6.16 among Soligas of Karnataka (Murthy,1987). Total Fertility Rate among various tribal groups is quite high in comparison to 3.8 (India National Population). Similarly, General Fertility Rate varies from a minimum of 125 among Jaunsaris (Kshatriya etal, 1997) to a maximum of 188 among Soligas of Karnataka (Murthy,1987). In other words, tribal population of India, in general, reveals high fertility.

It is unfortunate that Infant Mortality Rates for individual tribes are not readily available. It is very sensitive indicator of overall health culture and health behaviour of a tribe. There are many factors that determine Infant Mortality Rate like low birth weight, women's literacy, poor hygiene, poor sanitation, poverty.

By and large Infant Mortality Rate among Indian tribes varies from a minimum of 65/1000 live births among Sherpas (Gupta, 1980) to a maximum of 239/1000 live births among Gadabas of Andhra Pradesh (TRI, Hyderabad, 1980). Mean Infant Mortality Rate among Indian tribes was calculated to be 120.3 which is alarmingly higher than 77 for Indian National Population. Although infectious diseases account for the bulk of Infant Mortality Rate, but the role of genetic and genetico-environmental disorders becomes equally important because their frequency is quite high in Indian tribal groups.

Invariably, data on abortion and still birth are under reported. One of the reason could be their inability to share such sensitive information with the field workers who often stay there for a short period. And since most of us are non-clinicians, it becomes difficult to persuade the tribal people regarding the rationale of seeking such information. It takes time before people are convinced about the purpose of such information. This is just one of the several observations about the practical difficulties in obtaining information on such vital issues.

It is apparent that such data are available for few populations only, it can be seen that incidence of still births varies from, Zero to 7.4 per cent in Mishings of Assam.

A general indicator of the health of girls and women is their life expectancy. The expectation of life is the average number of years remaining to be lived by those surviving to that age. The expectation of life at birth is the life table function most frequently used as an index of the level of mortality. It also represents a summarization of the whole series of mortality rates for all ages combined as weighted by the life table stationary population.

Basu et al (1989), in their study on the Bastar tribal groups of Madhya Pradesh estimated the average life expectancy at birth based on q5 values for Muria – males

37.56 years, females 40.07; Maria – males 40.26 years, females 41.00 years; Bhatra – males 38.6 years, females 45.30 years and Halba – males 38.6 years and females 41.46 years; and as 41.1 years for all the four tribal groups combined. Although these figures are comparable to the rural non-tribal population of Madhya Pradesh, they were far below the average life expectancy at birth of 58.6 years for the Indian population at that time. Sex specific differentials were observed with males experiencing higher mortality than females. The differences can only be attributed to the prevailing socio-economic, cultural and health care practices of the four tribal groups.

Basu etal (1993) studied demographic features and health care practices in Dudh Kharia tribal population of Sundergarh district, Orissa. Using q5 values to estimate average life expectancy, it was observed that Kharia females show a high life expectancy (52.95 years) as compared to males (51.02 years). Sex ratio of Dudh Kharias (1101/1000) lends further support to the above observation. The general lifestyle of Dudh Kharias has been found to be relatively better than most of the surrounding tribal groups.

Health, Disease and Health Seeking Behaviour

Although the concept of well being and the notion of the disease varies between different tribal groups, yet, in tribal habitat, a person is usually considered to be afflicted with some disease if he/she is incapable of doing the routine work which is usually being expected to be carried out by that individual in the society, i.e., incapacitation from work is the universal index of poor health. Thus the concept of ill health becomes a functional one and not clinical (Bhasin, 1994). This is precisely the reason that among many tribal groups e.g. Kutia Kondha, Muria, Madia, Bhattra, Halba, Jaunsari, Santal, Lodha, Kharia, Bhil, Rathwa, Mina, Jatapu, Saroa, Pando, Khairwar,

Oraon, Munda and among many others; symptoms such as pains and aches, weakness, scabies, prolonged cough, mild fever, wounds etc. are not taken seriously as symptoms of disease.

A tribal in general, hardly makes a distinction in the magnitude of fever. However, within the limits of their own respective world view, most of the tribal societies have definite means for identifying and classifying various kinds of ailments and diseases. It may be worth while to state that at least one component of health is universally seen among the tribal societies, and that is, committing or omitting certain acts. In other words a breach of trust is thought to bring upon some kind of affliction on the individual or a family as a whole. Measles, tuberculosis, diarrhoea, cholera are some such diseases where individual's action may cause some concern to the family, clan or the village. Interestingly, the causation of such diseases is independent of the sanitary condition of the community/individuals. The fate of the individual and the community depends on their relationship with unseen forces which intervene in human affairs. If human beings offend them, the mystical power punish by causing sickness, death or other natural calamities. The tribal people believe in the presence of benevolent and malevolent spirits, the former playing a protective role, while the latter are considered to be responsible for causing disease and epidemics. It is a common observation across the tribal culture that the ancestral spirits play important role in the prosperity and protection of the family. And they have to be properly honoured for otherwise they will bring wrath on the family members. These spirits are believed to bring a state of physical, mental and social well being to the members of the family. There are many spirits who are feared because of their power to afflict people by bringing a number of diseases. The role of spirits, ghosts and dieties in the tribal life in the causation and treatment of disease is so important that the local tribal people have to seek the help of traditional diviners, medicine man, sorcerers (Sirha, Gunia, Bhua, Janteyar, Bhopa, Ojha, Pujari etc.) for appeasing, controlling or driving away the disease causing agents. The frequency of worshipping or getting in touch with malevolent spirits is more common because of their immediate effect on day to day life.

Rao et al (1954) have dealt with the problem of health and disease in Nainital. The authors observed that folkways centred around effect of image for cure of disease by shutting and preventing recourse to modern medicine. They also point out that indigenous medicines have the advantages in the context of folkways, the force of tradition, the conformation of mythology and legend, the prestige of strong personalities who act as functionaries, etc. They are interwoven with the common artifacts and techniques of daily existence and are easily available and inexpensive.

Carstairs (1955) in his study pointed out that the differences between the points of view of the physicians and the village folk with regard to the theory of etiology, techniques of curing and conceptions of roles of the physicians, resulted in misunderstanding between himself, a physician and the client.

Marriott (1955) conducted study on village Kishan Garhi in Aligarh district of Uttar Pradesh. This study tries to explore the problems of introducing western medicine in Indian village community. Marriott has shown how the contrasts and conflicts between the roles assumed by the indigenous and western medicinal practitioners resulted in obstacles to the acceptability of western medicine.

Elwin (1955) described the role of the Shaman among the Hill Saora of Orissa. The Shaman is one of the most important figures among the tribe. To the sick and lonely the Shaman is the nurse, friend, the guide, the analyst and the angel of strength and consolation.

Lewis (1958) noted that the advantage in learning about the indigenous beliefs and practices of the community is the insight they give into the world view of the people as concepts of disease causation are part of society's total world view, which is also reflected in other spheres such as agriculture, politics and interpersonal relations.

Dave (1960) conducted study on Garasia which revealed that, Garasia, like other tribal communities, have faith in indigenous herbs found in the forest around their habitat. In case of ailment, an elderly Garasia, who is known to have knowledge of various herbs, is consulted.

Opler (1963) lists the following as the most commonly believed causes of various diseases: (1) malfunctioning or imbalance of the three humans (Doshas) (2) faulty diet (3) lack of harmony with the supernatural world (4) activities of ghosts and witches (5) displeasure of deities (6) imbalance of forces which control health and (7) inappropriate behaviour in physical, social and economic matters.

Fuchs (1964) mentioned two types of medicine men – Jarka, who works through divination and Barwa, who works by calling to his aid a superhuman force. They and the villagers among whom they practice their art have a common faith in the techniques used.

Harper (1966) described a Shamanistic session in the Malnad region of Mysore. A shaman in this South Indian setting is a man who has a familiar spirit that he can ask to possess him whenever he desires. When he goes in trance, the spirit speaks through him.

Hasan (1967) distinguishes between two types of social and cultural factors that affect the health of any community: a) factors that directly affect the health of the community because certain customs, practices, beliefs, values and religious taboos etc create an environment that helps in the spread or control of certain disease and b) factors that indirectly affect the health of the community as they are related to the problem of medical care to the sick and the invalid. A physician who does not try to understand that cultural and intellectual level of the people and does not develop respect for cultural differences, proves to be a less successful practitioner in the village. If scientific medicine is to be carried to rural areas it is necessary for medical practitioners and public health workers to acquire adequate knowledge of the general concepts of culture and social organisation of that community.

Singh et al (1971) pointed out that any treatment involves two individuals, the healer and the patient, one requires the cooperation of both of them. Unfortunately, when a system of medicine has its roots in alien culture, there is conflict between different cultural values resulting in the rejection of the alien system by the recipient. The best way to overcome this problem is to understand the indigenous medical beliefs and practices and allow western medicine to find its place by following the cultural norms of the society in which it is to operate.

Jaggi (1973) have discussed how the nature of treatment varies with the type of cause identified. For example, religious rites occupy a prominent place in the treatment of diseases which are associated with supernatural causes to the exclusion of other factors.

Banerji (1982) discussed an unequal access to the health services between the social classes, the privileged and underprivileged and between rural and urban areas in the country. He pointed out that the frequent realization to take action to meet the health services need of the poor and to entrust "Peoples health in People's hand" indicates conceptual watershed to some significance in health services development in the country as this showed subordination of medical technology to the needs of the health.

Rizvi (1982) made study of the factors influencing health in a tribal community. The study in particular deals with socio-cultural analysis of Jaunsari community in U.P. It has been observed that climate and geography, beliefs, isolation and poverty influence attitudes with regard to health to a very great extent.

Gupta (1986) observed that Oraons and Ho's of Chottanagpur believe that disease is caused by breach of taboo, bad spirits, sorcery. They believe in spirit basis of disease. Besides magico-religious treatment the tribals have many natural remedies, decoction of forest herbs, of healing lotions known to the tribals. That means the tribal people have pharmacopia of their own for their many fold diseases like malaria, yaws, leprosy, veneral disease, bowel complaints, influenza, etc. Although there are block dispensaries and mobile hospitals, but the tribal people mainly depend for prevention and cure of their diseases on their traditional medicine-men. Gupta (1986) has also shown how the tribals conceptualize health, disease and remedy. The tribals have hit upon some herb in some way or other for the treatment of diseases.

Mann and Mann (1986) have observed that among Hakkipikki, a tribe of trappers and sellers in Karnataka is engrossed in the clutches of animistic beliefs. Damaging role of spirits, to health, is widely recognized. Supernaturals, for good and bad, are identified and the relevant treatment managed for. In addition to almost conditioning of social organisation by religion purview, the areas of health, both of individual and public are also experienced and explained in the deep realm of religious forces. They also have taboos and superstitions. Mann and Mann (1986) have mentioned that religion has an important place in tribal life. Various ingredients of tribal religion such as the gods, goddesses, spirits, ghosts, other allied and the associated rites and rituals react to diverse aspects of life including diseases, sickness and their cures. Mann and Mann (1986) have also mentioned that among tribes of western India, ill-health and cure practices are based on supernatural considerations. A strong feeling of the Bhil society is that the evil spirits haunt the recesses of the atmosphere and inflict harm to persons by causing various ailments and otherwise too. Mann and Mann (1986) through their field work experiences among the diverse categories of Indian tribes reveal that the inter play of spirit and man, with special reference to health, is of a far greater order among the tribals than one reported in case of the non-tribal rural communities.

Khan (1986) have tried to find out the impact of rural development programme on tribal health in Chottanagpur in Birbhum district. His study reveals that inspite of deep faith in traditional folk healing, the Santals are not apathetic to accept modern health practices, instead of ignoring their beliefs and culture, they are accepted with sympathy and trust. This informality in the inter-personal relations in the medical sphere motivate them to a great extent to accept modern health care and practices. The Santals in general like to follow traditional methods of delivery due to their poor economic condition. The traditional method of delivery is less expensive compared to delivery in health centre. This study depicts the failure of state Government health programmes in motivating the Santals.

Choudhuri et al (1986) studied the prevalence of diseases among the tribals in tribal concentrated areas of Purulia district of West Bengal and their attitude towards medicine. It was noted that the tribal patients having leprosy comprised of 23.6 percent of the total patients. Leprosy was considered as social stigma among the tribals. Tribals suffering from malaria and tuberculosis constituted of 24.0 percent and 27.5 percent respectively of total patients. Tribals were quite ignorant about the cause and characteristics of this disease. Village medicine man do not treat the patients suffering from these diseases. Allopathic doctors were preferred when tribals were unable to suppress the symptoms of the disease and in initial stages, tribal patients went to their own village medicine man. PHC was visited only if disease turns to an acute stage. Due to the distance factor district hospital is usually not visited by tribals for X-ray.

Bardhan et al (1989) conducted study among Bhils of Madhya Pradesh to know the utilization of modern health practices. It was found that non-tribal model of organizing and delivering health services is largely inappropriate for Bhils. There was general agreement that measures for health development for tribals and their delivery system cannot be the same as for the general population and there have to be variations (even inter and intra situation) depending upon the health status and consequent needs of the various tribal and non-tribal groups.

Chakravorty and Singhrol (1989) pointed out that indigenous medical system among tribes of Bastar could be divided into two groups namely ritual and herbal medicine. The tribal people are specialist in identifying, propitiating or controlling the evil-eyes which are generally believed by them to be the fundamental cause of most of the diseases. Diseases are also caused by unsatisfied souls of dead ancestors and names of dead persons are given to the grandchildren to convince the souls that they are remembered. They perform rituals, prayers and offerings to the deity. During seasonal festivals connected with economic life they perform dances and music to ward off diseases and natural calamities. In the second category of medicine-man are the herbalists who do not enchant mantras, though they use. Such herbalists are trained by their fathers and practice or familial profession.

Ali (1992) conducted study on tribal communities of Orissa. He found that an ecological imbalance caused by rapid deforestation has resulted not only in depleting food resources, but in prolonged droughts, adding to hunger and starvation.

Pacholi (1993) in his study on Bastar tribe of Madhya Pradesh found that tribal areas of Madhya Pradesh are the permanent source of malaria. Her study revealed that tribals blindly follow the local magic-man or the witch-doctor by different names such as 'Ladwa', 'Joshi-Panda' or `Gunia' in different parts of MP. The witch-doctor treats different diseases.

Basu et al (1994) conducted study on the perception of health and pattern of health seeking behaviour among the selected tribal population groups of Madhya Pradesh (Gonds, Muria, Madia, Abhuj Madia, Dhurva, Darla, Halbas and Bhattras) and Orissa (Kutia Kondhs and Kharias). The study revealed dismal picture as far as health status is concerned in terms of health seeking behaviour with the exception of some of the indicators in Dudh Kharia, though the trend is short-lived. Poor health seeking behaviour is attributed to social and living conditions, lack of competitive economy, malnutrition, illiteracy, ignorance, superstitions, improper and poor santiation, unsafe drinking water, unawareness of health education, under utilisation of health resources, cultural lag, poorly developed mass media, poor medical and health opportunities (variety, quality and quantity of health services) and infrastructure. It was also observed that there was great difficulty in persuading the tribal people to come to avail modern medical treatment because they have firm, rigid and well developed system of primitive medicine. They seek remedies through magico-religious practices to propitiate the supernatural powers. Basu et al (1994) studied the health seeking and health utilization pattern of Kutia Kondhs of Phulbani district, Orissa. The study revealed that "Jani" (the traditional healer) is the first person to be contacted in case of sickness. Kutia Kondhs consider modern medicine to be the last resort for seeking medical care and health care. The major reasons for low utilization of modern medicine were, the distant location of health care facilities, rough and hilly difficult terrain, poor economic condition, character and illiteracy, non-availability of medicines even from the health care institution etc. besides inflicting loss in terms of valuable working time for the activites related to livelihood.

Sachchidananda (1994) pointed out that a prerequisite to the success of any health programme in the tribal areas was to orient the doctors and all paramedical staff working in tribal areas to tribal life and culture so that they are able to communicate with the people on their own wavelength and to establish rapport with and instil confidence among them.

Singh (1994) examined the interplay of several factors influencing health of tribal communities namely (a) physical environment/elemental, forces of nature like earth, sun, wind, rain, forest, (b) Socio-economic state (c) Nutritional availability and dietary habits d) Psycho-social culture (e) Health culture and Health related behaviour (f) Mortality and Morbidity pattern (g) Genetic diseases and disorders, (h) The therapeutic system in vogue and (i) Health delivery systems. Singh pointed out that issues concerning tribal health, nutrition and gentico–environmental aspects are grossly under researched.

Sinha (1994) in his study on different tribal groups of India has discussed the scenario of primitive medicine, social and cultural activities being practiced by different tribal groups for the treatment of their diseases. Various types of herbal medicines, common names of the plants/ roots used by the tribals for the treatment of diseases had been enlisted by him. He has also discussed the development of modern allopathic system, health planning in India and the health situation of tribals.

Swain (1994) examined in depth the etiology of perception of health and perception of illness prevailing among the different tribal groups of India. It has been realised that diseases are not only due to physical, chemical or biological processes but also due to a number of socially and culturally determined factors.

Tandon (1994) discussed the wise variation in the dietary pattern of the aboriginal tribes of India living under contrasting conditions in different isolated regions. In tribes of North East Frontier Area, the dietary habits and other related modes of life are often suited to their best advantage and provides adequate nutrition and they may enjoy excellent health while in other tribes the diets are deficient, the energy intakes are very low.

Mahapatra (1994) analysed the Santal view of health and sickness, disease and death and also discussed the nature of linkage between traditional medicine and modern system of medicine and also the mechanism of exorcisms, spirit healing and medicine preparation.

Bhattacharya (1995) discussed the folk medicinal claims for the treatment of general diseases from Patalkot valley of Chhindwara district (M.P.) During the course of ethnobotanical investigations, some interesting, indigenous medicinal plants have been reported for the treatment of veneral diseases i.e., Gonorrhoea, Syphilis among the tribals (Bharia and Gond) of Patalkot valley, Eight indigenous plant species have been reported which cure different diseases. He has observed that in Patalkot valley mostly tribal men are used to practice the indigenous plants in their daily health care. In rural areas, veneral diseases are common among tribals due to unhygenic sexual practices. Tribals use indigenous plants for the treatment of veneral diseases.

Ghoshmaulik et al (1995) described the health situation of the Juangs, a primitive tribal group of Orissa. The disease prevalence among the Juangs suggest that these people are mostly victims of environmental infections. The study reveals that the Juang people are more dependent on the medicine man than the government primary health centre. The village indigenous medicine man or Shaman receives respect and faith, as the villagers think that he treats them well. The people visit the village medicine man more frequently because the primary health centres are located far away from most of the villages. It was found that the local health centres are not fully equipped with required medicines for the general treatment of the people of the villages.

Kapoor (1998) examined the concept of health, treatment and faith in a scheduled tribe of Himalayas and reported that the tribe has got more faith in traditional practices rather than modern. Kapoor (1999) reported the concept of disease and its healing aspect in various caste groups of Himalayas.

Mitra et al (1998) have examined the various environment, socio-economic and cultural factors whose interplay determines the health of pregnant and nursing mothers in 'Bhumiheen camp', a slum in Govindpuri, South

Delhi. They have tried to understand the health problems and practices of slum women and the role of socio-economic and physical environment in influencing them. They have found that the most direct effects of poor health and nutrition among females in countries like India are high mortality rates among young children and women of child bearing age and high morbidity rates throughout the life cycle. The effects of pervasive ill health extend beyond the women herself. Women's health and nutritional status is bound up with social, cultural and economic factors that influence all aspects of their lives, and it has consequences not only for the women themselves but also for the well-being of the family and functioning of households. For improving the health status of poor women, greater opportunities of schooling and employment as well as some changes in health system are needed, to increase the demand for health services and women's ability to use them. They have found that a perceived relation between poor surroundings and sickness, exists among slum women.

It has been observed that Health and treatment reflects the social solidarity of a community. In a tribal community, for example, illness and the consequent treatment is not always an individual or a familial affair, but the decision about the nature of treatment is taken at the community level. In the rural areas, in case of some specific diseases, not only the diseased person or his/her family, but the total village community is affected. All the other families in the village are expected to observe certain taboos or norms and food habits. The non-observance of such practices often calls for action by the village council. One cannot deny the impact of this psychological support in the context of treatment and cure. In some societies, illness may be an individual affair, and the patient may be looked after by the administration (state), while in tribal societies it is a familial or sometimes community affair where necessary action and care is taken by the community as a whole.

Among the tribals, there are a group of specialists-the priests and/or magicians or medicine-man whose services are sought after, depending on the cause of the illness. Among the Ho's (Roy,1922) there is the 'Deuri' or religious headman and 'Deonwa' or the spirit doctor. The 'Deuri' worships the deities and on occurrence of epidemics or diseases in the village, he offers a sacrifice at the sacred place. He is entrusted with the benevolent deities. The malevolent spirits are controlled by the 'Deonwa', often through magical performances. Among the Munda's (Majumdar,1937) and the Oraon's (Roy,1928) the religious headman is the 'pahan' who propitiates the village deities on behalf of the entire village to ward off disease and misfortune. Again, among some tribal groups, there is one person who officiates as priest and magician. The Bhils (Nath,1960) have 'Badava', who is the priest and performs the role of a medium, diviner, worshipper and healer. The Korwas (Majumdar,1961) beat an ailing person with iron chains and insert a burning wick into nose to cure him. When a Kuttia child of Orissa has its first serious illness it is generally decided that it is time for its naming ceremony to be performed. The medicine-man is called in for magical performances. Among the Kuttias (Ali,1980) the village priest is the magician. Tharu women (Vidyarthi,1963) of the Himalayas are pioneers in the magical arts. Dube (1970) has noted the strong belief of the tribals in invisible powers which help in controlling an epidemic and curing an ailing person. Buchs (1964) has noted two types of medicine-man: 'Jana', who works through divination and 'Barwa' who works by calling to his aid a superhuman force. The dependence on and confidence in traditional medicine-man or magicians and shamans are again often responsible for the non-acceptance of modern medicine. The traditional approach establishes faith and assurance in the patients, while modern medicine lacks this. As the shamans or what are called traditional medicine-man and magicians share the common cultural beliefs and practices of the patients, naturally they have more faith in them.

A few studies (Khan, 1986; Basu, 1994; Mahapatra, 1994; Sinha, 1994) have been conducted to study the interaction between traditional and modern system of medicine among the tribals.

There are some studies (Khan, 1986; Mahapatra, 1994; Sinha, 1994) which have shown that when both the facilities namely, traditional and modern, are available, the tribals often accept the modern system.

It has been noted by many that tribals living in remote areas have a better health status and more balanced food than those living in less remote areas (Chaudhuri,1986; Chaudhuri etal, 1989). Roy Burman (1990) has noted that physical well-being of the tribals is inextricably linked up with the forest ecology and rapid disappearance of forest, commercialization of minor forest produce, replacement of food crops by cash crops and tendency towards monoculture forestry and lastly privatization of communal lands may affect the health and nutrition and physical well-being of the tribal population.

Maternal and Child Health

Maternal and child care is an important aspect of health seeking behaviour. Maternal malnutrition leads to premature deliveries, low birth weight babies and pregnancy wastages. Data on this aspect is fragmentary on the tribal population groups of India. Nevertheless, the trends that are coming up are alarming. In one of the study on Bhotia, Kapoor and Prasad (1982) have observed that in the event of delayed conception of married women, the deities like Nanda Devi, Nabling and Kalika are propitiated by taking on earthen lamp in a plate for the whole night. This sort of appeasement for the procurement of child is known as 'Jagran' or 'Dardeo'. There is a popular belief among the Bhotias that a child born during the 'Amawasya' (dark phase of the moon) causes hardships to himself or his parents.

Prasad and Kapoor (1982) discussed the obstetrical practice among the Bhotia tribals. The Bhotias use very simple and unhygenic method to expell the placenta, locally known as 'Phali', from the mother's womb after child birth, if it is not expelled in natural course. The Bhotia women are also given a gentle abdominal massage and tea brewed from the 'Bodpa' leaves, found in Tibet, to effect easy delivery of the child.

In Bastar District (Basu and Kshatriya, 1989) it was observed that maternal and child care was largely neglected. Expectant mothers to a large extent are not inculcated against tetanus. From the inception of pregnancy to its termination, no specific nutritious diet is consumed by women. The consumption of iron, calcium and vitamins during pregnancy is poor. More than 90 per cent of deliveries are conducted at home, attended by elderly ladies of the household. No specific precautions are observed at the time of conducting deliveries which result in an increased susceptibility to various infections. Services of paramedical staff are secured only in difficult labour cases. It has been observed that a specially prepared concoction from various kinds of herbs known as "Kasa Pani" is provided to mother on the third day after delivery, "Kasa Pani" is a herbal medicine which according to them gives soothing effect, helps in recovering the depleting energy and also acts as anti-infective. Maternal mortality directly related to pregnancy and childbirth has been found to be appreciably high among the tribal populations of Bastar district. In addition, a lot of females suffer from ill health due to pregnancy and childbirth in the absence of well defined concept of health consciousness.

Devarapalli (1992) studied maternal care and obstetric practice among the Konda Dora tribes of Andhra Pradesh. It was found that Konda Dora women continued to work till the time of delivery. Dietary restrictions were also imposed on her during pregnancy. Beans were avoided as they were believed to cause epileptic fits. Deliveries were conducted at home. Umbilical cord was cut with the help of sterilized knife. After delivery the women remains completely at home for a minimum of one month.

Pandey and Tiwari (1993) carried out their studies on Hill Korwas of Madhya Pradesh. Of the 14 women, who were pregnant at the time of interview, only one had consulted ANM during her pregnancy. Only five seemed to knew about tetanus toxoid injections and intake of iron tablets during pregnancy. Most of the deliveries were conducted by untrained dais or traditional birth attendants (72 per cent) followed by elderly ladies (92.3 per cent) and less than one per cent were conducted by trained dais and the ANM. The cord was cut with a bamboo piece for a male child and with a knife for female child. The colostrum was not discarded. Goat milk was given as a substitute for mother's milk.

Pandey et al (1996) studied the maternal and child practices among the Khairwar women of Madhya Pradesh. The findings showed that pregnant women did not receive any special antenatal care or diet. There was no restriction on her daily work, including fetching of water and cutting of wood. The delivery was usually conducted by elderly women and the ANM or trained dai was not called and in case of complication the pregnant women was usually left at the mercy of god. The umbilical cord was usually cut with the help of sickle or old blade.

The chief causes of maternal mortality were found to be unhygienic and primitive practices for parturition. For example, it was observed that among Kutia Kondhs (Basu et al 1990), the delivery was conducted by the mother herself in a half squatting position holding a rope tied down from the roof of the hut. This helped her in applying pressure to deliver the child. In complicated labour, it might lead to maternal as well as child mortality. Similar crude birth practices were found to exist in other tribal groups like Kharias, Gonds, Santals etc.

As far as childcare is concerned, both rural and tribal illiterate mothers are observed to breast feed their babies. But most of them adopt harmful practices of discarding of colostrum, delaying the initiation of breast feeding and delaying the introduction of supplementary feeds. Vaccination and immunization of infants and children have been inadequate among tribal groups. Since the personal hygiene is very poor, the under 5 children are the worst sufferers and most vulnerable to infections. In addition, extremes of magico-religious beliefs and taboos tend to aggravate the problems.

Among the other morbid conditions, yaws, leprosy, tuberculosis and malaria have been found to be quite prevalent in tribal areas. Yaws is caused by Treponema pertenue and cases have been reported from (Bastar) Madhya Pradesh, Tripura and Maharastra (NICD 1993). Incidence of tuberculosis has been found to be increasing in tribal areas as the tribals are coming more in contact with other non-tribals. Poor nutritional status is also one of the contributory factors.

Incidence of leprosy has been reported in Scheduled Tribes of Bihar, Madhya Pradesh, Orissa, Uttar Pradesh, Tripura. Incidence of malaria is wide spread in almost all the tribal areas. Tribals who constitute 8.1 per cent of the total population contribute 25 per cent of the total malaria cases in the country. About 50 per cent of the P.falciparum cases reported in India come from tribal population. Malaria is a major health problem in all the tribal areas.

Very little information is available about the prevalence of non-cummunicable diseases in tribal population. Studies carried out by NIN (Rao et al, 1986) among the tribals of Sarguja, Jhabua and Bastar districts of Madhya Pradesh showed high incidence (53.9 per cent) of Goitre (enlarged thyroid gland) in tribals of Sarguja district. Eye conditions like cataract, corneal opacity etc. were found to be in high frequencies (20.3 per cent) in Bastar district.

OBJECTIVES

Health and culture are very much connected with ecology. The traditional health care system and treatment are based on deep observation and understanding of nature and environment. The hypothesis of the present study was to ascertain whether there is a interrelationship between socio-cultural practices, environmental conditions and health aspects exists among Saharia. The specific aims and objectives of the present study are:-

1. To study the ecological, demographic, cultural and economic aspects affecting the health of the tribe.
2. To understand the concept of health, illness and nature of treatment and to identify the pattern of health seeking behaviour.
3. To study the impact of various health programmes on the tribe and

4. To ascertain their preference for type of system of medicine and to identify factors associated with non-utilization of health services provided by government agencies.

Importance of the Study

This study aims to obtain an integrated picture of the community health and sickness, socio-economic condition, settlement pattern and environmental conditions against the background of the existing medical and sanitary facilities available in the primitive tribe Saharia. The Primitive Tribal Groups (PTGs) require greater insight and understanding than is required of other peoples and in this rather delicate and difficult talk administrators and field-workers ought to exhibit greater wariness, for we are dealing with a hoary culture which if owing to our inefficiency we cannot improve we should take care that we do not impair. While dealing with Primitive Tribes, what matters more is not the temporary displacement or loss of property but the damage caused to the cultural, psychological framework of the concerned people. It is not easy to pluck these people from their ancient moorings and press them into new grooves. The PTGs requires sensitive and delicate handling.

As there is no study on Saharia emphasizing on ecological, cultural and health aspect, this study has been carried out for improving their health and hygiene conditions.

Since there is paucity of comprehensive health related studies among the tribal population, therefore there is urgent need for initiating area specific, group specific, health need action research in consonance with the felt needs of the tribal communities. The present health oriented action research study will help in formulating need based health care strategy for the Saharia in particular and other tribes of Rajasthan in general.

Health is very intimately connected with ecology and culture. Therefore, this study will help in understanding of health problems and health care system of community using holistic and interactive approach.

2

Area and People

Area and People of Rajasthan

Rajasthan is situated in the north western part of the Indian Union (23^0 30' and 30° 11' North latitude and 69^0 29' ad 78° 17' East longitude). The total area of the state is 3,42,239 sq. km. and ranks second in terms of area in the country. It occupies 10.41 per cent of the country's total area yet its population is only 5 per cent of the total population of the country. The present state of Rajasthan is a union of 22 princely states and the centrally administered area of Ajmer and Merwara. The shape of the state is like an irregular rhomboid, covering a distance of 869 km. from west to east and 826 km. from north to south. The region to the west and north-west comprising of eleven districts spreading in 61.11 per cent of the total area is either desert or semi-desert which forms the Great Indian 'Thar' Desert. The Aravalli range of Hills–one of the oldest mountain ranges–runs through the heart of the state, extending to 69.2 kms, and dividing into two portions. The north-western portion is almost entirely a vast expanse of desert. On the other side, the south-eastern region has a varied terrain of extensive hill ranges, fertile table-land and dense forest. The climate of Rajasthan can be divided into four different seasons: pre-monsoon,

monsoon, post monsoon and winter. Pre-monsoon, extending from April to June, is the hottest with temperatures ranging from 32°C to 46°C. In the arid zone of west and north-west of Aravalli range, temperature often crosses 46°C, particularly in May and June. Post-monsoon is the second hot season with relatively uniform temperatures across the state. In October the average maximum temperature is 33°C to 38°C and the minimum is between 18°C and 20°C. The winter lasts from December to February. During this period the mean temperature ranges from minimum of 3°C to 25°C. The normal rainfall in Rajasthan varies from 253 mm to 1000 mm. The population of Rajasthan is 4.40 crores (Census of India, 1991) which is 5.20 per cent of the nation's population. Out of 44005990 persons, 23042780 are males and 20963210 females. The density of population in the state is 129 person per square kilometre compared to 273 for India. According to census 1991, 77.12 per cent of population in Rajasthan is rural and the rest 22.58 per cent is urban. The literacy rate (for the population aged 7 years and above), according to 1991 census, was 39 per cent in the state compared to 52 per cent in the country. The parent language of Rajasthan is Avanti. Rajasthani (Khadi-boli) is the lingua franca in Rajasthan. Scheduled Castes and Scheduled Tribes form 17.29 per cent and 12.44 per cent, respectively of the total population of Rajasthan. About 8 per cent of the Rajasthan's population is Muslim, most of whom, are Sunnis. There is a small affluent community of Shia Muslims in south-eastern Rajasthan, known as Bohras. Rajasthan is the land where all religions have flourished and co-existed in complete harmony for centuries. Religion has always played a important role in the lives of the people of Rajasthan. Hindus form the majority in the state, their customs, manners etc. are same as in other parts of the country. Folklores are also worshiped over here, along with all the forms of nature each area has its own local deity. The important religious centres are Pushkar, Nathdwara, Deshnok, Amer etc. Jainism has also flourished for centuries in Rajasthan. The Jains have some of the most exquisite temples in Ranakpur, Mount Abu, Shri Mahaveerji, Osian, Jaisalmer, Chittaurgarh and other parts of the state. There are many important centres of Muslim pilgrimage in Rajasthan. The most important being the dargah of Khwaja Moinuddin Chisti at Ajmer, it is regarded as one of the greatest sufi shrines of the world. The other important centre is the Atarki Dargah, the tomb of Hamiuddin Nagori at Nagaur. Sikhs and Christians and other communities reside in the state, giving Rajasthan a cosmopolitan flavour. Population of Districts of Rajasthan and sex ratio and density are given in tables 5 and 6 respectively.

Table 5: Population of Districts

District	*Total Population (1991)*
1	2
RAJASTHAN	43 880 640
Jaipur	4 719 257
Udaipur	2 885 039
Ganganagar	2 618 914
Alwar	2 286 701
Nagaur	2 137 258
Jodhpur	2 127 552
Kota & Baran	2 027 375
Sawai Madhipur	1 953 807
Sikar	1 836 572
Ajmer	1 723 081
Bharatpur	1 646 501
Bhilwara	1 591 236
Jhunjhunun	1 565 488
Churu	1 539 470
Pali	1 484 691

1	2
Chittaurgarh	1 482 267
Barmer	1 433 351
Bikaner	1 209 107
Banswara	1 154 964
Jalor	1 141 604
Tonk	973 118
Jhalwar	955 510
Dungarpur	874 329
Bundi	768 150
Dholpur	748 326
Sirohi	653 324
Jaisalmer	343 648

Note: Baran has been made District in 1996. Earlier it came under District Kota.

Table 6: Sex Ratio and Density of Population 1991

District	*Sex Ratio (females per 1000 males)*	*Density (persons per sq. km.)*
	1991	1991
RAJASTHAN	910	129
Ganganagar	878	127
Bikaner	887	44
Churu	940	91
Jhunjhunun	949	264
Alwar	889	273
Bharatpur	835	325
Dholpur	796	247
Sawai Madhopur	857	186
Jaipur	982	335
Sikar	952	238
Ajmer	924	203

Tonk	925	135
Jaisalmer	810	9
Jodhpur	904	93
Nagaur	949	121
Pali	957	120
Barmer	891	50
Jalor	942	107
Sirohi	950	127
Bhilwara	946	152
Udaipur	966	167
Chittaurgarh	950	137
Dungarpur	997	232
Banswara	969	229
Bundi	891	138
Kota & Baran	888	163
Jhalawar	918	154

TRIBES IN RAJASTHAN

According to the 1991 census, the tribal population in Rajasthan is 54,74,881, out of which 28,37,014 are males and 26,37,867 are females. It is nearly 8.10 per cent of the total tribal population of the country and 12.44 per cent of the total population of Rajasthan. Of this tribal population, 9.35 per cent live in rural areas and the rest 4.65 per cent in cities. The Scheduled Tribes Amendment Order,1976 has given the following list of twelve tribes in Rajasthan:

1. Bhil, Bhil-Garasia, Dholi Bhil, Dungari Bhil, Dungari Garasia, Mewasi Bhil, Rawal Bhil, Tadvi Bhil, Bhagalia Bhil, Pawara, Vasava, Vasave.
2. Bhil-Meena.
3. Damor, Damaria

4. Dhanka, Tadvi, Tetaria, Valvi
5. Garasia (excluding Rajput Garasia)
6. Kathodi, Katkari, Dhor Kathodi, Dhor Katkari, Son, Kathodi, Son Katkari
7. Kokna, Kokni, Kukna
8. Koli Dhor, Tokre Koli, Kolcha, Kolgha
9. Meena
10. Naikada, Nayaka, Chotiwala Naika, Kapadia Naika, Mota Nayaka, Nana
11. Naika.
12. Patelia
13. Seharia, Seharia, Sahriya, Saharia (A Primitive Tribe). The present study was conducted on Saharia of Baran District of Rajasthan.

Regarding geographical distribution of tribals, Rajasthan state can be classified into three regions (Southern region, South-Eastern Region and Western Region).

(i) **Southern Region** : It has the maximum tribal concentration (59.9 per cent of the total tribal population of the state) in Rajasthan. The Districts of this region are Udaipur, Dungarpur, Banswara, Chittorgarh, Sirohi, Pali, Bhilwara and Jhalawar with a tribal population of about 36.79 per cent (10,63,071), 65.84 per cent (5,75,805), 73.47 per cent (8,49,059), 20.28 for cent (3,00,971), 23.39 per cent (1,53,005), 5.39 per cent (80,265), 9.26 per cent (1,43,748) and 11.89 per cent (1,13,834) of the total population of these districts respectively (Table 7). The main tribes of this region include Bhil, Bhil-Meena, Damor, Garasia and Kathodi.

(ii) **South-Eastern Region**: It has a tribal concentration of about 32.12 per cent of the total tribal population of the state. The major Districts are Alwar, Bharatpur, Sawai Madhopur, Kota and Baran, Bundi, Jaipur and Tonk with a tribal population of 8.05 per cent (1,85,048), 2.31 per cent (38,212), 22.58 per cent (cent (4,43,469), 14.19 per cent (2,88,367), 20.25 per cent (1,56,001), 11.25 per cent (5,31,652) and 11.89 per cent (1,15,948) of the total population of these districts respectively (Table 7). The important tribes of this region are Meena, Saharia, Kathodi, Patelia, Naika and Koli.

(iii) **Western Region**: It has only 7.98 per cent of the total tribal population of the state. The Districts include Barmer, Jalor, Jodhpur and Sikar with a tribal population about 5.86 per cent (84,232), 8.43 per cent (96,324), 2.82 per cent (60,819) and 2.64 per cent (48,667) of the total population of these Districts respectively (Table 7). The tribes of this region are Dhanka, Kokna and Naika. The tribal population in Rajasthan is thus concentrated in belts running from Sirohi through the Udaipur, Dungarpur, Chittorgarh and Banswara districts to the Bundi, Kota & Baran, Sawai Madhopur, Tonk and Jaipur districts. The borders of Banswara, Dungarpur, Udaipur and Sirohi districts meet the Sabarkantha, Banaskantha and Panchmahal districts of Gujarat State, while the Saharia populated Shahabad and Kishanganj tehsils of Baran district share a common boundary with Madhya Pradesh.

According to 1991 census, there are 28,37,014 tribal males (12.31%) and 26,37,867 females (12.58%) of the total male and female population of the state respectively. The

Table 7: Tribal Population in Rajasthan

S.No	Name of District	Total Population	Total Tribal Population	% of Total Population of the District	% to Tribal Population of the District
1	Jaipur	4722551	531652	11.25	9.71
2	Udaipur	2889301	1063071	36.79	19.41
3	Banswara	1155600	849059	73.47	15.50
4	Dungarpur	874549	575805	65.84	10.51
5	Sirohi	654029	153005	23.39	2.79
6	S.Madhopur	1963246	443469	22.58	8.08
7	Chittorgarh	1484190	300971	20.27	5.49
8	Bundi	770248	156001	20.25	3.84
9	Kota&Baran*	2030831	288367	14.19	5.26
10	Tonk	975006	115948	11.89	3.11
11	Jhalawar	956971	113834	11.89	2.07
12	Bharatpur	1651584	38212	2.31	0.69
13	Bhilwara	1552584	143748	9.26	2.62
14	Jalor	1142563	96324	8.43	1.53
15	Alwar	2296580	185048	8.05	3.37
16	Barmer	1435222	84232	5.86	1.46
17	Pali	1486432	80265	5.39	0.08
18	Jaisalmer	344517	16697	4.84	0.30
19	Dholpur	749471	34429	4.59	0.62
20	Sikar	1842914	48667	2.64	0.89
21	Ajmer	1729207	39764	2.29	0.72
22	Jhunjhunu	1582421	30528	1.92	0.55
23	Churu	1543211	7819	0.50	0.14
24	Ganganagar	2622777	8945	0.34	0.16
25	Jodhpur	2153483	60819	2.82	1.11
26	Nagor	2144810	4796	0.22	0.11
27	Bikaner	1211140	3185	0.26	0.05
	Rajasthan	44005990	5474881	12.44	

Source: Rajasthan Census Report (Scheduled Tribes) 1991.
*Earlier Baran was a town of District Kota. In 1996 Baran was made a separate District.

gender ratio among tribals was 929 females per 1,000 males which exhibits a decline in female population as compared to 945 in 1981 census. Literacy among the tribals in Rajasthan is also very low. According to 1991 census, only 19.44 per cent of the tribal people are literate, of which 33.29 per cent are males and 4.42 per cent females. The present study was conducted on Saharia of District Baran.

DISTRICT BARAN

District BARAN, the place is situated at a distance of 72 km by road and 67 km by train from the District Kota. It is said that it was founded by Solanki Rajputs during 14th or 15th century. The place derives its present name because it is believed to have been populated by the inhabitants of twelve (Barah) adjacent villages. The Saharia are the only Primitive Tribe of Rajasthan and inhabit in the Shahabad and Kishanganj tehsils of Baran District. The total area of Baran District is 2898 sq. km. Shahabad tehsil has a total area of 1469 sq. km whereas Kishanganj tehsil has an area of 1429 sq. km. The number of villages in Shahabad tehsil are 236, out of which only 146 are inhabited villages and rest of the villages are uninhabited. The total population in Shahabad tehsil is 83028. The population of Saharia in Shahabad tehsil is 28632 which forms 34.48 per cent of the total population. The population of Saharia in Kishanganj tehsil is 19190 which forms 17.71 per cent of the total population. The number of villages in Kishanganj tehsil are 203, out of which only 171 are inhabited villages. The District Baran contributed 1.84 per cent to the total population of the State. Density of population of Rajasthan works out to be 129 sq. km whereas it is only 116 persons per sq. km in case of Baran District. To cater to the needs of Saharia, special programmes were started in, 1977-78 for the development of Saharia residing in Kishanganj and

Shahabad Panchayat Samitis of Baran District (formerly Kota District). According to Saharia Development Project Area (1991), the total population of Saharia in Baran District is 47,822 (Table 8).

Table 8: Details of Saharia Development Project

District	*Name of Panchayat Samiti*	*Area in sq. km.*	*No. of Villages*	*Population as per 1991 Census*				
				Total Population	*S.T. Population including Saharia*	*Saharia Population*	*% of S.T. with Total*	*% of Saharis with Total*
Baran	Shaha-bad	1469	236	83028	28829	28632	34.72	34.48
	Kishan-ganj	1429	203	108345	37032	19190	34.18	17.71
	Total	2898	439	191373	65861	47822	34.41	24.99

The Saharia of Rajasthan are concentrated in two tehsils of Shahbad and Kishanganj which also coincide with their respective Panchayat Samitis. The Shahabad and Kishanganj tehsil can be divided into two distinct divisions (1) Upreti or Upland and (2) Taleti or Lowland. The Upreti constitutes Kishanganj teshil and the western part of Shahabad tehsil. The remaining part of Shahabad tehsil comes under the classification of Taleti.

WORK DONE ON SAHARIA

It has been observed that very few studies have been carried out among Saharia (Russel & Hiralal, 1916; Census of India 1961; Gupta, 1965; Bhil, 1973; Vyas & Bhil, 1978; TRI, 1984; Naik 1984; Ghatak & Nanda, 1985; Dalat, 1987; Shekhawat, 1987; Kapoor & Kshatriya, 2001) focusing on ethnographic account, economic aspects, religion and selection intensity.

Gupta (1965:1, 24, 28, 34) in his study of Sanwara village has stated "Saharia are reckoned to be the most backward, hence the poorest of the scheduled tribes of Rajasthan, were brought and sold like chattels till half a century ago. Though the maximum contribution to the village income is made by the Saharia, their number being the largest, their average income is the lowest. That each of the twenty families (out of thirty–two) all Saharia, having an annual income of under Rs. 360, are as good as slaves of their money-lenders, who have become grafters and occupy key positions in the Panchayats and Party Organisations, and wield considerable influence in the area. The poor illiterate folk cannot escape their clutches, and have even to repay time-barred loans. The Saharia had been economically so subservient to the Ahir that no love was lost between them, and the latter exploited the depressed section so thoroughly that for a mere cupful of curds the Saharia was made to work the whole day". Considering the Saharia as untouchable, they are not permitted to share the crematorium and drinking water well meant for caste Hindus. Gupta (1965:4) has mentioned " What to speak of taking water from a common source, when an Ahir woman carrying water comes across a Saharia woman, the latter is expected to move to one side leaving the way clear for the former".

Most of the time the Saharia have to approach Bohras (local money-lenders, traders and exploiters) to manage for bullocks, seeds and implements. The Bohras also fed them for lean months. But all these transactions involve a serious element of exploitation. Still more needy Saharia get into the system of "Batoli-Hali" and "Hali-Bata" with the Bohras and other landlords of the area. These systems also speak of bondage and exploitation. Under the system of "Batoli-hali", a landlord engages a Saharia at a petty amount per month to plough his land. But in reality the man engaged is not given the amount. His remuneration is made adjusted against the quantity of grain which the person borrowed for his family during the lean or scarcity months. The man keeps on repeating it since the scarcity is almost normal feature every year. The man who gets the grain to eat during scarcity months remains so much obliged that he hardly dares to ask any amount. The same arrangement is made when a loan, in cash, is needed. The landlord advances the loan and gets the man pledged for his services. He does so as a Hali, the ploughman. The man serves for the stipulated period against a petty amount. For rupees to three hundred, a man serves for the year. The man also gets some quantity of grain while working with the landlord. Though the needy is helped during the moment of crisis, the relationship involves a strong element of exploitation. Major victims of this are the landless ones, or even those who own small and uneconomic holdings. In "Hali-Bata" system a man is engaged as Hali, the ploughman, on the condition that he would get one-fifth to one-sixth share of the agriculture produce. At the same time the landlord gives him one pair of dress made of coarse cloth. He is also given one seer of coarse grain. But for the family members of Hali the landlord does not give anything free. The food grain and cloths given to them are treated as loan at very difficult terms and conditions. For the whole year the accumulation of interest on these items becomes so bulk that it is adjusted against the one-fifth to one-sixth share which the Hali was supposed to get. As a result the man pledged does not, in practice, receive any share from the landlord. The cycle goes on confining the man to bondage. The haves create such a condition that the have-nots have no easy escape from their clutches. Writing on Hali-Bata system of Saharia, Gupta (1965:19) further states "For financing agricultural operations, borrowings in kind is very common in the area and it is called Hali-Bata. The money- lenders advance the required quantity of grain for the maintenance of the tiller's family on the understanding of sharing the produce equally. After harvest the quantities advanced are taken into consideration while fixing the share. By this arrangement a Saharia family of four to five persons share for six-to-seven hours of daily work extending over four months."

Regarding Saharia caste Panchayat Gupta (1965:35) states " The jurisdiction of Saharia caste Panchayat extends not only over the villages of Rajasthan but also over the villages located in Madhya Pradesh. It is more or less like a central counsel of elders for the tribe, invested with quasi-judicial powers. It is composed of twelve members called Patels, all equal in status. The headman of caste Panchayat had government backing in old days. That backing being now removed, there is proportionate decrease in their effectiveness".

Analysing the life of Saharia in Census Monograph Sanawara, Gupta (1965: 37) mentions that "Even the independence in 1947, and the consequential integration of the princely states in the region and the guarantee of equality enshrined in our constitution gave them but little protection from suppression, social demands and grinding economic exploitation. Total illiteracy, chronic poverty, indebtedness and ignorance have made them lethargic and indifferent to their fate. A number of ameliorative measures have, no doubt, been adopted for their uplift during the first two plan periods in the field of education, co-operation and rehabilitation. In 1956, the Government of Rajasthan introduced a scheme providing for allotment of twenty Bighas of land to each Saharia family with subsidies for construction of wells and houses, and purchase bullocks. A plan for setting up model villages was also introduced, as a result of which three villages viz.,Shubhghara, Khusalpura and Ghagattaghatti, all in the vicinity of Sanwara, were remodelled. The results, however, were not encouraging. The truth is that the Saharia of Sanwara are caught in a vicious circle of poverty.

Saxena (1976) has thrown light on the aspects of situational structure, social organization especially related to marriage, kinship system and clan (gotra) organisation, economic structure, religious beliefs, law and justice system besides the facts about the problems of Saharia.

Tiwari and Pandey (1989) described the economic status of Saharia to be very low. According to them this tribe's base was in the forests in the past. With the cleaning of forests they have been forced to depend on simple jobs as labours. Uneducated untrained, unskilled and mostly all the Saharia people are below the poverty line. Besides being unorganised they have been reduced to merely agricultural labourers, that too the bonded labour, the worst form of agricultural labourer those are two types of bonded labour namely "Hali Batholi" or "Maheendar". They only get food and flour and in the return, they have to work for the whole year in other form of bonded labours "Hali Banta" in which they work as daily wagers they are paid as bonus at the end of the year, which generally coincides with the harvesting season. Though the Social Welfare Department ventured to colonise some sections of Saharia , it has not made much of impact on the economic status. Exercises went on randomly without taking stock of Saharia life and culture (Mann, 1989). Saharia suffer from lack of infrastructural facilities and are, to a great extent, not freed from bondage. Welfare measures have generally proved ineffective as the unplanned forces, money lenders and land lords, neutralised them.

The Saharia are not so powerful magicians like Baiga. Their magic of to-day is weak and gradually they are losing faith in the efficacy of magicoreligious rites. The magician among Saharia is known as 'Janteyar'. The office of 'Janteyar' is not hereditary and anybody can become 'Janteyar' by learning the secret of this art. The 'Janteyar' stands between mankind and all the malignant forces of nature and super-nature, his only weapons are a fruit stick (Brown) of straw and a winnowing fan. The Saharia also believe in the malevolent activities of the witches. A witch is known as 'Toni', 'Churel' or 'Dakan'. It is commonly believed that a woman who dies in pregnancy or during child birth or is eaten away by wild animals generally become a witch and she is a source of constant trouble to others. Whenever a witch hunts somebody, 'Janteyar' is consulted to suggest the remedy. The recitation of some "Mantras" is often effective to defeat the nefarious activities of witches.

Pandey (1989) also mentioned that Saharia tribal people wear simple dresses. The men wear half size "Dhoti" and typical shirts, popularly called "Saluka". They also wear the "Pagadi"(turban) to cover their heads. The Saharia women also wear simple dresses. They either wear "Ghaghara" or "Lungra" sort of petticoat below the wase and "Choli"- a bra like blouse. Saharia women are also fond of silver ornaments. They wear silver necklace and silver bangles in both hands. When the Saharia people visit outside their homes, they generally wear the best dress, they can afford. The Saharia man would wear surely his pagadi and the Saharia women, the silver ornaments.

3

Methodology

Anthropological studies are characterized by their holistic approach of study based on first hand information collected through fieldwork. In fact field is the laboratory for anthropologists. Fieldwork tradition in anthropology is a recent development, but fieldwork has become an integral part of anthropological studies. Fieldwork is the method of data collection which provides an anthropologist with first hand data based on first hand and intimate contact with the people under study. To understand the life of a people or to study a particular human society, a knowledge of their environment, their history and the extent to which contact with other people has taken place is also necessary. To collect first hand information on all these aspects anthropologists engage themselves in fieldwork. An important characteristic feature of anthropological enquiries is that they are holistic in approach. An anthropologist may select a particular problem for study but tries to see it with regard to its place in the total milieu and the life of the human group concerned (Firth, 1951). Investigation in the field of social science, is a means to an end and not an end in itself. These investigations are usually undertaken for various purpose such as administrative, legislative, political, economic, commercial, industrial and scientific. The purpose influences the nature and scope of investigations. In every aspect of human life practical knowledge is more valuable than theoretical. To study the methodology of field investigation, practical work in the field is more effective than the theoretical one. Before going to actual field it is essential that one should equip himself both intellectually and materially. The intellectual equipments are language, study of people and study of allied objects, whereas material equipments are provision for habitat, dress, first-aid medicines and other comfort articles, mechanical aids including photography.

One of the main styles of social research used by empirically oriented social scientists is field research; a style of investigation that is also referred to as 'field- work', 'qualitative method', 'interpretative research', 'case study method', and 'ethnography'. This approach to social investigation has traditionally been associated with social anthropologists, whose `field' consisted of a small-scale society where it was possible to do `research' by living and working among the people. Gullick indicates "when the anthropologist is in the field, field work is his total life. He copes with it by using his whole body and personality in the same way that he copes with life when he is not in the field............life in the field involves the same emotions as life at home: elation, boredom, embarrassment, contentment, anger, joy, anxiety and so on. To these are added however, the necessity of being continually on the alert (of not taking one's surroundings and relationships for granted), and the necessity of learning new routines and cues. These necessities are likely to force a heightened awareness of facets of one's personality of which one had not been aware before. This can be an emotionally devastating experience, but it is by no means inevitably so" (Gullick, 1977: 90). In this respect, field research is learning situation in which researchers have to understand their own actions and activities as well as those of the people they are studying. The main instrument of social investigation is the researcher, who has to learn the local language, live among the people and participate in their activities over relatively long periods of time in order to acquire a detailed understanding of the situation under study. Field research involves observing and analysing real-life situations, of studying actions and activities as they occur. The field researcher, therefore, relies upon learning first hand about a people, a culture. However, if the researcher is to obtain an insider's view of situations, it is vital to maintain an outsider's perspective (Powdermaker, 1966). Field researcher, therefore have to develop self-criticism and self-awareness, if involvement and detachment are to be achieved in social situations. In this respect, researchers maintain membership in the culture in which they were reared while establishing membership in the groups which they are studying; they are socialized into another culture. This has been commented upon by Evans-Pritchard as "Perhaps it would be better to say that one lives in two different worlds of thought at the same time, in categories and concepts and values that often cannot easily be reconciled. One becomes a sort of double marginal man, alienated from both worlds" (Evans-Pritchard, 1973:2-3).

Malinowski (1922) considered that ethnographic material was only of value when it was possible to distinguish between direct observation, native sentiments and interpretations, and the inferences of the author. It was vital that some assessment could be made concerning an author's acquaintance with facts and the conditions under which observations were made. It was Malinowski who raised the question about how an ethnographer should work. He considered that ethnographers needed to know, the aims of their studies, to live among the natives without other Europeans and to collect data by means of specific methods. In particular, he argued that ethnographers should cut themselves off from other Europeans and live among the natives as this was the only way to gain some appreciation of the social processes involved in everyday life, and to get to know individuals, their customs and their beliefs. Malinowski also considered that studying natives in their natural setting was preferable to using paid informants. This was basic to all field research.

In anthropological fieldwork the conception and the approach derive from the fact that the problem is essentially human in nature. That is why honesty of purpose is really important, because the anthropologist is observed far more intensely by the people than he can ever observe them, and a false step is soon noted. That is why sensitivity is also an important factor. Sometimes when an opposition to his investigation develops, an alertness of the equipments of the people, an ability to give way peacefully at the right moment and a politeness and tactfulness can ease a difficult situation. Here a sense of humour is of great value. An anthropologist should talk with full affection to his friends amongst the tribal people with whom he is to work. So honesty, sensitiveness and an abundant sense of humour are essential for a field-worker in approaching tribal people in the field. If he is equipped with these, he will find the horizon of the field clear in spite of unavoidable difficulties and frustrations in his path. A fieldwork in an alien illiterate society entails the combination and improvisation of a wide variety of research techniques in order to render a holistic picture of the community and its social phenomena.

Pelto (1978) writes that "the field worker is the principal research instrument" the objective data collected is filtered through the subjective perceptionary states of the researcher differing for various fieldworkers. Also they differ with respect to the events and scenes that capture their interest and appeal to their predilections. Also the sociological paradigm with which a researcher identifies and views the social world of the community apart from being conditioned by a sound theoretical knowledge, defines problem of study, methodology embraced and is again an index of his/her personality and predilections.

Social anthropology studies, social customs, values, social institutions and the way in which these are interrelated. In other words it focuses on the study of man in a social aspect, i.e., in his relationship with other people in higher communities i.e., such relationships are enacted by man as the member of society in accordance with the culturally prescribed norms and values. But societies are not "things" in any material sense. They cannot be set up in labs to enable experimentation or any kind of study of man under controlled conditions. To quote John Beatie "The concept of society is a rational and not a substantial one, the only concrete entity's given in a social situation are the people" (1964). Thus social anthropology aiming at studying human relations have to take the society as a laboratory. Their hypothesis about social and cultural institutions and their inter connections have to be tested in the course of fieldwork in the societies and in natural conditions.

Methodology refers to the structure of procedures and transformational rules guiding data collection (Pelto, 1978). Bailey (1994) calls it the philosophy of research process. According to Pelto (1978) methodology is "the logic in use involved in selecting particular (research) observational techniques assessing their yield of data and relating these data to theoretical propositions". Since the research techniques or concrete methods employed as the "pragmatics of primary data collection" dictated by the field contingencies are only logical and practical they are a part and parcel of the broad methodology adopted. There is a large range of methodological techniques for collecting different kinds of information, which technique a researcher chooses depends upon the appropriateness of any particular technique for any specific topic being studied, the logistical difficulties of applying the methods under consideration, and the researcher's theoretical views about the validity of various techniques.

The research process can be made inter subjectively verifiable by gauging the people's interpretations, the true meanings underlying their local terms, the symbolic significance attached to their world making it comprehensible. Further all people's inferences of a phenomenon, even if contradictory, within the same community, are to be reported. The emic interpretations once gauged, a juxtaposition of the etic analysis stemming from the exegetical framework may be undertaken to render the work academically useful.

A methodology deals with ethnographic research. Ethnology is comparative, drawing materials for analysis from a wide variety of societies around the world. These materials for analysis come originally from anthropological work i.e., from ethnography. Literally, the world "ethnography" means to write about peoples. As we use the term, it refers to the descriptive study of human societies. Ethnography provides the building blocks for cultural anthropology (Hoebel, 1972). Field methodology has its roots in the preposition laid down by Malinowski (1922). The genesis of this approach is traced back to Malinowski's (1922) pioneering fieldwork among the Trobriand Islanders. Malinowski is seen both as the pioneer of modern anthropological fieldwork carried out through participant observation and as one of the founding father's of modern social anthropology. There is one well established, basic strategy that is common ground among anthropologists since Malinowski (1922) set the standard early in the twentieth century. That strategy is participant observation', long-term residence among the population being studied. It is generally agreed that ethnographic field work should be carried out for a year or more, that the researcher should be able to converse effectively, preferably in the local language, with the people being studied, and that some kind of engagement with the flow of ordinary life and ongoing activities of local people in the research setting is desirable.

A researcher's methodological approach is influenced by her or his vision of the goals of social or cultural anthropology and by heuristic theoretical position. Two visions of objectives and goals are well established in contemporary social and cultural anthropology. These visions might be labelled 'humanistic' and 'scientific', and are perhaps better seen as poles around which researchers situate themselves, and between which they shift in one direction or another from time to time, thus shifting somewhat the tendency of the entire discipline. In the scientific vision, the goal of anthropology is the discovery of descriptive generalizations and explanatory laws about the way society and culture work which can account for the commonalties and variations among societies and their trajectories over time. To accomplish this, attention must be given to behavior as well as ideas, and precise information is required, quantitative as well as qualitative. To maximize the value of the information, and to limit errors resulting from human subjectivity and bias, systematic forms of data collection are needed, and checks, as in repeat studies, are required. Comparative studies juxtaposing different societies and cultures allow the formulation of general explanations. The methodologies favoured by humanistic anthropologists would include the use of empathy, as in participation in local activities to gain a feeling for local life, the collection through casual conversation, interviews and life-histories of expressions by local people of their perspectives, understandings, and experiences, and the recording of collective expressions of local culture such as myths, proverbs, songs, stories, rituals and ceremonies. The methodologies favoured by anthropologists of a scientific leaning would include general surveys, observations of behaviour patterns, tests and other formal exercises of information collection, structured interviews, sampling of the population studied and comparison of different neighbourhoods, communities or cultural groups. The intention is to use methodological techniques to collect systematic and precise information on the relevant variables so as to test decisively the hypothesis, thus adding to firmly established knowledge about social and cultural patterns and the nature of human life.

Pilot Study

The pilot study is an exploratory study done before the actual work is started in the field. It is a pre-testing of the research methods and techniques in order to perfect them. It makes one aware of difficulties beforehand and the required modification of the techniques are made to suit field conditions. Use of a Pilot can lend credence to the ability to manage qualitative research by describing initial observations or interviews. "First days in the field" experiences usually reveal fascinating questions and intriguing patterns. A description of initial observations demonstrates not only the ability to manage this research, but also the strength of the approach for revealing enticing research questions. Inclusion of a description of a pilot study or initial observations can strengthen the proposal.

Prior to the actual fieldwork, a pilot study was done for six weeks (June-July 1999). The primary objective was to acquire a basic idea of the villages and thus prepare a mental framework as how to proceed with my work, and also to establish a formal rapport, to identify some key informants. This was very important to make them know my presence and objective, among the Saharia. Instead of questioning them while they were wholly unaware of my motive. It is always advisable to take precaution during rapport establishment, as the slightest fault committed in any innocence may cause the subjects to become entirely hostile. On my introductory visit to the Saharia villages, rapport establishment began on the way itself.

SITE AND SAMPLE SELECTION

The selection of the village for study has important bearings on the generalizations of the results on tribal life. The Shahabad and Kishanganj tehsils of District Baran are divided into Upreti (Upland) and Taleti (Lowland) areas. For the present study I selected eight villages from Taleti (Shahabad, Kusalpura, Faredua, Devri, Beelkheradang, Chaurakhadi, Pathari and Sandhri) and eight villages from Upreti (Shahpur, Gangapura, Khandasehrol, Ghesua, Dhikwani, Hatri, Mahodara and Samrania). Before I decided to select villages for the present study, I visited a number of villages of the Shahabad and Kishanganj tehsils and contacted Mr. Badan Singh Verma (Area Organiser, Adim Jati Sevak Sangh, Shahabad) and obtained general information such as the total population, number of households, literacy etc. in the villages. The data was collected from 300 households (150 households of Taleti and 150 households of Upreti).

Rapport Establishment

Rapport establishment for an anthropologist includes forming a good relationship with the people, knowing the people telling them about oneself and one's work etc. A properly conducted introduction to the head of the community or an influential person helps immensely in getting public support. An amiable manner is a necessity, coupled with polite language and a sober dress. Small gifts may also help, though no false promises should be given. Establishing rapport helps the people of the community to trust the researcher. The extensive writing of anthropologist Rosalie Wax (1971) has emphasized the importance of the researcher's initial contacts with the members of the society or group chosen for study. The reciprocal relationship between the host and the field-worker enables the latter to avoid foolish, insulting, and potentially dangerous behaviour, to make valuable contacts, and to understand the acceptance and repayment of obligations. "The most egregious error that a field-worker can commit" (p.47), according to Wax, is assuming that tolerance by hosts also implies their high regard and inclusion. Researcher should plan and practice role maintenance. Research design demonstrates the researcher's awareness about managing impressions that facilitate the research. They provide evidence that they know enough about the setting and the people, their routines, and their environments to anticipate the effect of the researcher's presence. Researchers then advise roles that elicit cooperation, trust, openness, and acceptance. At times this means play-acting, dressing a certain way, and allowing themselves to be manipulates. Frequently, researcher acquires multiple roles in order to develop research relationships with different people. Researchers benefit from careful planning of their roles—since most participants reject insincere, manipulating people. In addition, researchers have to teach the participants what the researcher`s role is. They must acquaint participants with the activities that the role involves, with the sorts of information that fall within the purview of the study, with the possible uses of the information, and the manner in which the participants could aid in the research. The researcher cannot be simply a sponge-like observer who just soaks up data, as Vidich (1963) explains, since many people will not respond to or trust someone who will not take a stand. Roles and relationship do emerge in a field. But one must demonstrate in the proposal a logical plan and time for easing of roles.

The process of choosing a field site is followed by an attempt to gain an entry and acceptance by the people. A good rapport involves reciprocal friendly relations between the natives and the field worker who has to abide by the cultural norms and prescriptions and accept the contextually defined rationality of their folk beliefs and practices. "I began to feel that I was indeed in touch with the natives and this certainly is the preliminary condition of being able to carry out a successful fieldwork" (Malinowski, 1922). A rapport thus entails maintaining a contact, an effective communication with the natives and is the key to a fruitful field experience.

In the first two months of my fieldwork, I made a detailed survey of the village, going to the village in the morning and coming back in the evening. During this period, in the beginning I found that the Saharia were suspicious towards outsiders. When I interviewed people regarding the diseases prevalent in the village and what they do when they fall ill some of them questioned me like this: "Babuji, Ka aap doctor ho, meaning: "Sir, are you a doctor?" Some others said, "Saab ka aap humko dava doge", meaning: "Sir, can you give us medicines". Some persons start narrating to me their problems and plight. During my talk with the village folk, I tried to explain my purpose and told them that I was a student from Delhi University and as part of my study, I will enquire about the problems of health and disease in the villages and then write a book on it. I told them further that this book would be examined by my teacher and if he approved of it, I would then get the degree from the university. I always tried to be uniform in my explanations so as to avoid confusion. The researcher cannot, therefore, escape from explaining the purpose of his work in the village. It is not necessary to tell them everything . I do not agree with Edward Gross (1956) who said that student should not only explain his purpose enough, but also, "what he says, must be true". In

Primitive Tribal Groups (PTGs) of India it would make no sense to say that one is an anthropologist. Confidence came gradually after the people knew me well enough to have decided that I could be trusted. My continued presence in the village and participation in day to day activities helped me considerably in gaining the confidence of Saharia. Thus while conducting a survey of the villages for two months by going there in the morning and coming back in the evening, I finally pitched up an abode in the village. A room in the school was given to me. As the days went on I came in closer and still closer contact with the local people. I could memorise their names and gain personal friendship of many of them. While visiting houses, I paid special attention to children. I often talked to them in a humorous mood, started playing with them for a while in the presence of their parents and sometimes distributed sweets and biscuits among them. This helped me considerably in coming close to their parents and families in the village.

I kept a packet of bidis with me and distributed them among Saharia whenever we gathered for gossip. Sometimes, I smoked funnel-shaped clay pipe and cigarette with them. I also attended marriage parties in the villages. But achievement of mutual understanding takes place within the field of restricted possibilities and the researcher's choice of participation is limited by the range of role adaptations he is capable of as well as those which local men under study find acceptable (Paul, 1953). For example, although I smoked bidi, cigarette and chillam once or twice with some villagers, yet I could neither drink with them nor could I smoke since I was opposed to smoking and drinking.

The first two months of my fieldwork went into the surveying of the villages of Upreti and Taleti. I spent the first ten days in establishing rapport with the people making friends with them explaining to them my purpose of arrival and stay in the village. Since I introduced myself to the villager's as a university student from Delhi to see the village, the people seemed to be very enthusiastic to talk to me as I was from Delhi. This made it easier for me to approach the people. On my part I tried not to miss a single opportunity to establish a friendly relation with the people.

In order to be friendly with the children and to make them feel at ease with me, I distributed sweets, biscuits and toffees among them. The friendship established with the children and the response of the children indicates my rapport among them. The school children were very eager to get themselves photographed when they saw the camera in my hand. So, sometimes I had to click the camera in order to please them. The villagers also reciprocated quite satisfactorily. They invited me for tea. Whenever I used to go to any family for the purpose of collecting some information they would hardly let me go without offering tea. Everywhere I was given the treatment of a special guest.

The role that an anthropologist holds in the field plays an important role not only in rapport establishment but it has its affect also on the consequent phases of data collection. In order to establish a good rapport the role of the anthropologist should be such that it is acceptable to the people and their culture. My role in the village was that of a student whose humble intention was to know the village, its people and their way of life. The beginning of fieldwork was marked by several instances when people would come and narrate to me their problems and plight. Some of the boys and young man became very good friends of mine. They were usually free in the evening and then they would sit with me for a few hours at a stretch, answering all my queries to the best of their knowledge. Some of these boys and young men became my key informants, Inder, Dinesh and Pappu who were studying in the village school told me about their school, the teachers and their fellow students. They also told about what they do at their homes and the short stories prevalent in the village. The other folks too expressed happiness about my coming to the village and staying with them. They always treated me with love and affection and responded well to my queries. Shiv Shanker Sharma, Block Supervisor of PHC Shahabad helped me to have a deeper insight into the village life. He invited me to his house also and treated me with tea and snacks. Some girls and the young women of the village became friendly and were not hesitant in talking to me. The teacher of the village school treated me with regard. He always expressed his ready willingness to help me in my study. His behaviour allowed me to approach without any hesitation during the school hours also. Kiran Dogra, Project officer (Lok Jumbish Pariyojna) was very modest and helpful. She answered all my queries in a very friendly way. She provided information about literacy rate among the Saharia. She also told about the life of Saharia. Neeraj Sharma, agricultural supervisor (Shahabad) invited me to his house for lunch. He also helped me in collecting secondary data.

In the villages when I talked to the Saharia suffering from T.B, fever, headache and other diseases and the treatment resorted to by the natives this seem to arouse curiosity in the mind of Saharia and they asked me, "Do you have a cure for diseases such as T.B, fever, headache, bodyache etc". I told them that I have received no medical training and was carrying only simple medicine for minor aches and pains. Moreover, I told them that I had to know more about their diseases to be able to help. Few Saharia told me, "We are suffering from headaches and because of it we are not able to sleep in the night and we don't have medicine". Since I was carrying a strip of Saridon tablets,

I gave them tablets of Saridon and told them to have it only when they have pain, with water and after the meals. They asked me to come the following day, probably wanting to test its effectivity. The next day they greeted me with thanks and told me that the medicine had relieved them of the pain and enabled them to have a good sleep. The T.B patients then narrated to me their illness, the advanced stage of T.B for which they had resorted to a catalogue of treatments indicating medical pluralism. They had lost faith in the traditional healer 'Janteyar' and were currently on allopathic treatment. Some patients showed me the tablets they got from the TB control camp. The interviews highlighted the abject poverty of their existence and the coping strategies as well as the alarming unhygienic conditions with practically no avoidance from the family members. I advised the T.B patients to continue with allopathic treatment and attend the T.B control camp every month and to practice avoidance because the disease (T.B) being infectious.

STUDY DESIGN

In the study design, the researcher must anticipate questions about the credibility and trustworthiness of the findings. Poor sampling decisions may threaten these findings. To justify a sample, one must know the universe and all of its relevant variables—an impossible task. Generally, the best compromise is to include a sample with the widest possible range of variation in the phenomenon, settings, or people under study (Dobbert, 1982).

The following different aspects were taken into consideration for the study:

A. ECOLOGY: Physical Environment

(i) Geographical, geological and climatic, landscape of the area under study.

(ii) Land, hills, rivers, terrace, rainfall, temperature, humidity etc.

(iii) Flora and Fauna.

(iv) Settlement Pattern

B. CULTURE

(i) Technology, economy and property inheritance.

(ii) Marriage, family, kinship and religion.

(iii) Food production

(iv) Technological development

(v) Education

C. HEALTH

(i) Tribal concept of health

(ii) Cultural practices, environmental and economic factors affecting health

(iii) Interaction of traditional and modern medicine systems

(iv) Health care delivery systems

(v) Role of health service providers

(vi) Allopathic health providers

(vii) View of medical officer with regard to tribal health need

(viii) Health worker

(ix) Non allopathic service provider

(a) Great tradition healers

(b) View of tradition healers

(x) Private practitioners

(xi) Views relating to practitioners with regard to tribal health

(xii) Health seeking behavior and health care utilization among tribals – concept of illness, health seeking behavior, attitude towards traditional healer and felt need of tribal population with regard to health services, attitudes of tribals towards medical services provided at PHC, attitude towards family planning.

(xiii) Health Programmes of State and Central Government and their impact on tribals.

(xiv) Health care services, accessibility, utilization pattern and broad constraints.

DATA COLLECTION

Rapport establishment also has another primary function. It facilitates to a great extent the collection of data from the community. This data is of two types: Primary data and secondary data.

Primary data is that which the investigator or the researcher collects himself. Secondary Data is collected from reference sources like the library, etc. The second method is used for comparative methods-Anthropologists, generally prefer to work with primary data. There are two main techniques of data collection. These are Intensive fieldwork methods and Survey methods.

Intensive fieldwork methods include observation, interview, case study and genealogy. After the preliminary stage of rapport establishment, the actual process of data collection began. I began the study by conducting a household survey of the villages. It included basic information about each household, viz, the total number of members, name of the head, tribe, clan, age, sex, education, occupation, marital status. Besides it included information regarding land ownership. Data regarding the various aspects of village life was collected in a very informal way. Talking to the villagers in a very casual manner, I tried to gather as much information as I could. I also asked the villagers about the general village life depending upon their interests and moods. For any specific information, I tried to approach the individuals associated to that field. For example to know about the schools, I talked to teachers and students. For finding about the medical facilities, I talked to Doctors and ANM working in the PHC. For knowing about the traditional medical system I talked to the 'Janteyar' (Faith Healer). A lot of information was gathered and verified by discussion with different informants. Information was also cross-checked by asking the same questions from different members of the village.

TECHNIQUES EMPLOYED IN DATA COLLECTION

In order to find out the relationship between ecology, culture and health among the Saharia both primary and secondary data was collected.

Primary Data: Both qualitative and quantitative approaches were followed for the collection of primary data.

Quantitative data: Quantitative study was designed to generate information with the help of a set of schedules. A household schedule was prepared consisting of closed, open and multiple response questions. To confirm the responses in the schedule, the same question was asked in different ways two or three times during the interviews/ triangulation.

Household Profile

Information from the households was obtained regarding:

- Demography
- Morbidity

- Maternal and Child Health
- Family Planning
- Heath utilization and awareness

Village Profile

Village profile schedule was developed to obtain general profile of the village. The village profile schedule was used to obtain description of the community household structure, sources of electricity, availability of important public health measure and educational facility.

Qualitative Data

In order to generate free flow of information on various issues of tribal health both from tribal respondents and service providers the following approaches were used.

Various qualitative techniques include Focus Group Discussion, In-depth interviews, Observation, Case Studies, and Narratives.

Observation: Goode and Hatt (1952:119) remarked, "Observation may take many forms and is at once the most primitive and the most modern of research techniques. It includes the most casual uncontrolled experience as well as the most exact film records of laboratory experimentation". Observation can be more detailed observation or it can include participant observation. The value of participant observation as a tool for collecting data has been recognised from fairly early in the study of small scale societies. This is because the disadvantage of the investigator being an outside agent is minimized. I used the method of participant and non-participant observation. Observation is systematic viewing of phenomenon with a definite purpose. Observation may be controlled or uncontrolled depending on the investigator's control over the situation in which a particular event or activity is observed. Again depending on the degree of participation by the investigator in the life and activities of the native people, observation may be categorised as participant and non-participant observation. There were certain types of information that were obtained through direct observation because observation helped in the recording of behaviour at the time of its spontaneous occurrence. It often provides facts which best local informants can not even think of reporting. It is also not uncommon that there are events in the life of the people about which people want to maintain secracy and which they are not willing to discuss with others, particularly outsiders. For example, in the villages people are usually not prepared to name those who frequently break taboos. But if one is present in the village, one can easily spot such offenders. One can easily see drunkards in the community, even though the people are reluctant to talk about them. Hence in certain circumstances observation demands less active cooperation from the subjects, although it can not overcome resistance totally.

Participant Observation: In participant observation, the investigator takes active part in the event being studied. The observer becomes a member of the community by adopting various roles. Since the distance between the observer and the observed is narrowed down, participant observation becomes more of an experience marked with some degree of nativization.

Anthropologists of the statute of Evans-Pritchard (1950), Firth (1957), and Malinowski (1953), have all contributed to a better understanding of the societies studied by them through participant observation. Taking of field work technique mention must be made of Malinowski's classic, 'The Argonauts of the Western Pacific' and Firth's (1957) study of the Tikopia. Kaberry (1963:71) says Malinowski a compliment he richly deserves by saying, "Malinowski as the chronic and spokesman of the Trobrianders gave ethnography dimension it had hitherto lacked, actually of relationships and richers of content".

My study, therefore, is the result of extensive and intensive fieldwork carried out during a period of 8 months (Sep. 1999 April 2000).

NON-PARTICIPANT OBSERVATION

The investigator observers the events and activities from outside, since the investigator here does not participate in the event or become a part of the society he may fail to get the native's view of an event or activity. Due to these limitations of these two methods investigators mostly use Quasi-participant observation i.e. observation of participation and personal involvement is decided and controlled by the researcher himself. Observation, in whichever form is used is the principal among all data collection techniques. "Not only do they contribute to the basic, varied stock of knowledge about social relations with which we all begin our study, but they are principal data-gathering techniques for many modern investigations" (Goode and Hatt: 1952). During field work I used this technique not only to collect information about the village in general and the people, their customs etc. but also to collect information relevant to my topic of study.

Interviews: There are several kinds of information that can not be obtained by direct observation and the researcher has no choice but to ask people. For example, I was interested in knowing what the people of the village did in the case of an snake–bite in the village. Had I insisted on observation, I should have waited till the time an snake–bite case occur since there was no single snake – bite case during the period of my stay in the village. The best way of collecting such information was naturally to ask people direct questions in this regard instead of wasting time otherwise. In the same way, observation can not help if the researcher is interested in knowing past events which might throw light on the present. In short, he will have to interview people for such purposes. Hence I used interview and observation as supplementary methods. An interview is a method of data collection that may be described as an interaction involving the interviewer and the interviewee, the purpose of which is to obtain valid and reliable information.

In the initial stage of data collection and my introduction to the people I made use of informal interviews. Later on, after necessary rapport establishment, more formal and structured interviews were applied. An interview guide was also prepared for the purpose of conducting interview so that the points that I was intending to cover in the interview could be properly discussed. Sometimes while interviewing I tried to note down some important points. Whenever I felt that my writing down of the information made the informant uneasy, I listened to the information more intently and later on used to write down the information in isolation to avoid any confusion later.

The interview method is much favoured for collecting information. It is a particularly useful tool when coupled with observation. Madge (1958:50) says, "The interview.................and its half brother, the questionnaire................ is popularly regarded the method par-excellence of social science". Since a great majority of rural population in India is still illiterate, the researcher can not use questionnaires in such a society. What he can do to find out beliefs, motivations, anticipations, future plans or past events, is to interview them. I used structured interviews. Although the questions and their order was pre-determined by me yet I made my interviews flexible. I put often supplementary questions to my respondents. But for few, the questions were framed in such a manner that they simply raised an issue but did not provide any structure for the respondent's reply. Its aim was to give maximum opportunity to answer in their own frame of reference and terms.

IN-DEPTH INTERVIEWING

In-depth interviewing is a data collection technique relied on quite extensively by qualitative researchers. It is often described as "a conversation with a purpose" (Kahn & Cannell, 1957:149). I used In-depth interviews also. In-depth interviews were carried out with health providers in order to obtain their views about health status of tribal population. In-depth interviews were carried out with different health personnel of the block as mentioned below. A total of twenty-one in-depth interviews were carried out. Information was obtained with the help of pre-designed check list.

Designation of Health Personnel	*No. of Persons Interviewed*
BMO (Block Medical Officer)	1
MO (Medical Officer)	1
HWF (Health Worker Male)	6
HWF (Health Worker Female)	6
Private Practitioner	3
Traditional Healer	4

CASE STUDY

The approach used to explore and analyse the life of a social unit, be it a person or a faculty or an institution or a community or even a culture is known as a case study. It is a way of organizing social data so as to preserve the unitary character of the social unit and of the objects being studied. In this holistic approach, history is also accounted for. The events are recorded as the individuals see it, emphasizing on its history, development and social interaction. Since this means that entire process of the case has to be studied therefore it becomes a processual study. A case study is a situation whose participation in the making of the whole and the interaction with it are studied. Extended case studies of patients involving the clinical interviews, focussing on the health aspects of the individual's personal life history, together with a follow up of observed cases were useful in illuminating the native perceptions of etiological classical of diseases, their narration and interpretation of their symptomatic illness behaviour, the decision making role, healer-patient interaction, mode of diagnosis by the specialist, curative practice and medical pluralism in terms of the treatment resorted to, as a part of therapy management for severe ailments. The latter may entail an abandonment of the ethnomedical therapy or its combination with cosmopolitan medicine ; Illness viewed as a process.

Sixteen case studies were carried out to cover the following topics:

- Awareness and treatment pattern
- Local beliefs and practices
- Accessibility factors such as cost, time involved, availability of services and medicine, attitude and behaviour of the medical staff and traditional healer.

FOCUS GROUP DISCUSSION (FGD)

A focus group session is a discussion in which a small number of respondents usually between 6-12 members, under the guidance of a moderator, talk about topics that are believed to be of special importance to the investigator. In recent years, there has been a notable growth in the use of focus group session research to gain insight into the dynamic relationships of attitudes, opinions, motivations, concerns and problems related to current and projected human activity (Calder, 1977, Mcdonald, 1979; Cage, 1980). Participants are chosen from some specific target groups whose opinions and ideas are germane to the investigation, usually more than one group session is conducted to ensure adequate coverage. They capitalise on group dynamics and allow a small group of respondents to be guided by a skilled moderator into increasing levels of focus and depth on the key issues of a research topic (Feyisetan, 1994). The discussion is the basis from which information is obtained. It is conducted as an open conversation in which each participant may comment, ask questions to other participants, or respond to comments by others. Interaction among the respondents is encouraged to stimulate in-depth discussion of various topics. FGD can be used as a primary data collection method for some topics that cannot be easily studied through quantitative methods. FGD are particularly suited to subjects that are of a sensitive or personal nature, eg., Sexual behaviour, abortion and FP practice. It could be used to find out what motivates people to use a specific health production service facility. Such background information can be critical to health planners who need to know how the population views various health issues. It can also be used to pinpoint problems and to generate idea for service delivery improvements.

Focus group session research does not seek to statistically quantity group norms, traits and characteristics but to expose the underlying attitude and opinions. The quality of the response is important and the purpose is to detect directions of behaviour rather than magnitude. Therefore sessions are rigorous in the depth of the inquiry. As a result of dynamic group interaction focus group discussion sessions offer a number of advantages. Given the proper environment, participants are less on guard against personal disclosures because the atmosphere is friendly, tolerant and permissive. Personal revelations are facilitated by other group members who support, comment or disagree as well as by the relative homogeneity of the group which acts as an encouragement for all to express strong opinions or ideas. The lively dialogue activates memories, feelings and experiences in a member similar to the process of free association. The relatively free format of the FGD allows the moderator to pursue unexpected avenues which are relevant to the topic at hand but could not have been foreseen beforehand. Twelve Focus Group Discussion (Six each in Upreti and Taleti villages) were carried out to cover the following topics.

1. Knowledge, attitude and practices (KAP) with regard to health.
2. Health seeking behaviour of tribals, awareness of family planning, awareness about ANC (antenatal care) check-up, factors restraining women from utilizing ANC services, consumption of food during pregnancy and child immunization etc. Guideline was prepared on each topic. Focus Group Discussion was conducted by researcher (moderator) and a panchayat member acted as a recorder. Recorder also noted down non-verbal expression of focus group members when they were at discussion.

Secondary Data: The information concerning the past significant events in the District Baran (formerly Kota District) in state of Rajasthan was obtained from various official and non-official reports including the District Gazetter, Kota (1991) and reports on Census of India. The secondary data was also collected from office of Saharia Development Project (Shahabad), Adim Jati Sevak Sangh (Shahabad),Office of Collector (Baran), Asefa Project

(Baran), Information Centre (Kota), Lok Jumbish Pariyojna (Shahabad), Sankalp Sanstha (Mamoni). Several books were consulted in Central Science Library (CSL),Central Reference Library (CRL),National Medical Library (NML), National Institute of Health and Family Welfare (NIHFW), National Social Science Documentation Centre (NASSDOC), National Council of Educational Research and Training (NCERT), Ratan Tata Library (RTL), British Council Library(BGL), Library of South-Campus.

PROBLEMS ENCOUNTERED DURING THE COURSE OF FIELD WORK

To carry on his work an anthropologist observes village folk, visiting their huts, listening to their tales, joining in their social and religious rites and observing their behaviour, manners and customs. Usually an anthropologist carries on field studies among the people who are outside the main historical stream. Often these people live in far-off places, particularly in the hills and jungles, where they eke out a miserable existence; and to reach them an anthropologist may on occasions have to face many difficulties such as physical dangers, diseases, lack of transport, food, etc. But this aspect of the anthropologist's work is covered by romanticism and the spirit of adventure and thus has an appeal for the public; it has been consequently often over-stressed. But the real fieldwork is a scientific discipline. This makes vigorous demands on the patience, perseverance and sense of humour. An anthropologist studies the people among whom he lives and works, because from them he can gather data that will throw light on the main problems of his science. In Argonauts of The Western Pacific (1922), Malinowski also showed that he was aware of some of the problems associated with field research. In particular, the questions surrounding the impact of the observer on the observed and the influence of the observer upon village life were examined. However, these issues were considered unproblematic as Malinowski claimed that the constant presence of the anthropologist ceased to be a disturbing influence upon tribal life. Indeed, Malinowski advocated participation on the part of the researcher, as he remarked "In this type of work, it is good for the Ethnographer sometimes to put aside camera, note book and pencil, and to join in himself in what is going on. He can take part in the native's games, he can follow them on their visits and walks, sit down and listen and share in their conversations (Malinonwski, 1922: 21). Such participation, it was argued, allowed the researcher to obtain understanding of the lives of the people studied. I found Saharia to be too curious to know about me. They thought I am from city and highly educated. Saharia thought that I am a rich person. I usually avoided moving in the village wearing European dress or especially Jeans or T-Shirts or goggles, which I used, in the city. While conducting the fieldwork, I appeared before the Saharia in a simple shirt and trousers. This was the most comfortable dress in the villages, as one could sit with peasants in the field, on grass and other places of work without being afraid of spoiling one's clothes. Moreover, the villagers did not consider this dress to be something extraordinary. In fact, some of them also used it occasionally.

The field researcher needs to guard his/her dressing and behaviour pattern what Berreman refers to as "impression management". The dress should be simple, unattractive and subtle, the behavioural pattern in conformity with the norms and local standards. In the fieldwork situation of a traditional village setting I observed that a researcher needs to be especially careful in his/her dressing and behaviour to avoid being categorized. This categorization stems from the exposure of the natives to the cine world of the city. Although posters of professional actors and actresses adorn the walls of some households the native attitude towards them is one of derision and contempt owing to their "bad", a term loosely implying vulgar profession. Thus when a young researcher encroach upon their culture, his/her differing dress and behaviour tend to invite native value judgements an understanding of which reveal the native perceptions on "good" and "bad" behaviour. Few minor behavioural indulgences labelling a male as "bad" involve engaging in loud talk and laughing openly with females, wearing of sunglasses, a pretentious attitude and conspicuous use of cosmetics and perfumes. The young researcher thus has to confirm to minor behavioural expectations considered appropriate in the context, and has to resocialise one self. While conducting the interviews in the village, I found it difficult, if not impossible, to talk to the respondents alone in isolation. Usually neighbours also sat down and sometimes passerby stopped and listened to the proceedings. The least that happened was that family members only were present. It was not uncommon for family members, neighbours and passerby to interfere, or answer the questions before the selected respondent had a chance to answer himself. Obviously it would have done no good, had I got annoyed at this every time. To meet this problem, I tried to make it clear to others in a friendly manner that I needed one person's answers and I would be interviewing many others also. I always tried to ascertain the view of the selected respondents in case some one else answered before he got a chance to do so himself. While interviewing a TB patient in the interior village, I came across a critical situation when his relatives and neighbours got angry and told me that he is telling lies. They said that he is foolish. Then I chaged the topic.

Interviewing women was not free from difficulties. Talking to women on sensitive issues related to their perceptions concerning maternal and child health, sexual diseases and pregnancy disorders, the cure they resorted and the role of midwife (dai) created suspicion in their minds which enforced me switch to some other topics, yet it sensitized me to the health care practices among women tilting my paradigm to dwell more upon the gender related status attribution in each sphere and absorbing the women's world view. The 'Janteyar' (faith headers) often wondered on my curiously obsessive need to learn about their esoteric magico-religious domain rather than focussing on topics obvious as agriculture, education, economic organization etc. For this I had to reiterate that my topic of study on various kinds of diseases, and its perceptions as embedded in their magico-religious domain left me with no choice. I told them that I had a fascinating interest in knowing about the phenomenon of mediumistic divination as a healing system almost absent in my own culture. Further I used to express my respect for their beliefs and practices and my desire to write on it for it to be revealed to a larger audience in the academic sphere. This served to reinforce their sense of pride in their sacred cultural apparatus and usually elicited an enthusiastic response. Sometimes the Saharia also asked my views regarding certain diseases. A Saharia patient asked " do the medicines prescribed by doctors at PHC has some side-effects"? I told him that the doctor might know it better. I am not a doctor. I did not comment on it because if I started educating them it might give me a different picture. I was more interested in knowing their views.

Inspite of my best efforts to collect the maximum possible information, data collection was not smooth all the time. Some kind of discrepancies crept in between-such as prejudice of the informant about my intention etc. For instance while interviewing a person, I had to ask my informant about various members of the family, their occupation, marital status etc. which created suspicion in the minds of some of the members of the family. The wife of my informant was so suspicious that she indirectly told her husband not to disclose anything to me as she thought that I would disclose their information to some higher authority and might cause trouble to them. Barring a few exceptional cases like this the overall process of data collection was interesting and at the same time enjoyable.

The folk are generally busy with their work and what to an anthropologist is a significant interview might not be looked upon by the peasant as productive. However it requires tact and patience and one should wait for a suitable time to interview an individual. For this context of presenting credentials and being accepted special mention should be made of Neeraj, agriculture supervisor at Shahabad, who initially smoothered the way at first by allaying the doubts of my informants regarding my bonafides. He told them about the nature of my work and that my questions were directed towards academic approach. The interview sessions with an informant were incumbent upon his mood and time available to him. The faith healer ('Janteyar') got mentally fatigued while disclosing information on the esoteric aspects. Mukundilal frequently ended the interviews abruptly, asking me to come again. Language did not pose much of problem owing to the prevalent 'khariboli' in the Saharia belt. However slight language strain was faced in conversations with Saharia of the interior villages. Problem or recall of exact ages was faced in the present investigation e.g. age at birth, age at death and age of women and men in other non-literate societies without birth and death certification. However, health workers of PHC who maintain age information of tribal people, who were primary source of age information also. To further authenticate this information other religious and social events were also taken into consideration. Most of the individuals remembered these events. Poor awareness of tribals made it difficult getting spontaneous response or information regarding health utilization pattern especially of preventive services. Exact annual income of tribals couldn't be assessed as they did not have fixed earning throughout the year. Their earning pattern varied according to season and availability of different job with varied wages.

Lastly in the concluding note, I would like to express my gratitude and point out that without the co-operation of Saharia this fieldwork would not have been a success. The time for departure had come and I said good buy to the village. With fond memories in my heart, I left the village in a happy mood without any bitter experiences.

DEMOGRAPHIC VARIABLES

In the present study few demographic variables have also been analysed:

1. **Measures of Age Sex Composition:** A comprehensive Interview Schedule is prepared pertaining to questions relevant to the topic. A household profile schedule was also prepared to obtain information related to age, sex, head etc. of the households.
2. **Sex Ratio**: Sex ratio is the number of females per thousand males.

$$\text{Sex Ratio} = \frac{PF}{PM}$$

where PF = Total population of females
PM = Total population of males

3. **Masculinity Proportion**: Masculinity is the measurer of sex composition. The formula for the masculinity proportion is:

$$\frac{Pm}{Pt} \text{ X } 100$$

where Pm represent the number of males and Pt the total population

4. **Crude Birth Rate (CBR)**: It is defined as the number of births per thousand population in a given year.

$$CBR = \frac{\text{No. of Births in a year}}{\text{Total Population}}$$

5. **General Fertility Rate (GFR)**: It is the "number of live births per 1000 women in the reproductive age-group (15-49 years) in a given year".

$$GFR = \frac{\text{No.of live births in an area during the year}}{\text{Mid year female population age 15-49 in the same area in same year}}$$

6. **Age Specific Fertility Rate (ASFR):** It is defined as the number of live births to a mother of a specified age-group per thousand mid-year female population of the same age group. The age-specific fertility rates throw light on the fertility pattern. They are also sensitive indicators of FP achievement.

$$ASFR = \frac{\text{No.of live births to mothers of a specified age group}}{\text{Mid-year female population of the same age group}} \times 1000$$

7. **Total Fertility Rate (TFR)**: Summation of ASFRs over all ages. As the ASFRs are frequently calculated in conventional quinquennial age groups, the sum is multiplied by 5. It is generally expressed as per women.

ASFR X 5 = TFR

8. **Crude Death Rate (CDR):** It is defined as the number of deaths per 1000 population in a given year.

$$CDR = \frac{\text{No. of Deaths in a year}}{\text{Total Population}}$$

9. **Infant Mortality Rate (IMR):** It is the number of deaths to infants under one year of age in a given year per 1000 live births in that year.

$$IMR = \frac{\text{No.of Deaths aged less than 12 months}}{\text{No. of live birth occurred during a year}}$$

Analysis has been made on Health Seeking Behaviour and Maternal and Child Health Care (MCH) of the community.

4

Socio-Cultural Life of Saharia

History

The word Saharia appears to have been derived from the Persian World, "Sehr"meaning jungle. The Muslim rulers reckoned Saharia as inhabitants of forest. According to the Census monograph (1961) 'SANWARA' the word Saharia is derived from Persian word 'Seher' meaning desert. The occupation of the region, along the river Indus, by foreign rulers seems to have compelled them to quit the area. These are possible conjunctures and need historical support. From etymological point of view 'Saharia' means a companion of tiger (SA=Companion, and HARIA= tiger in the riddle verse of saint Surdas). "This clearly shows that Saharia has always been a forest dweller living in the midst of jungles and because of his age-old dependence on forest he is so bold and never afraid of moving in the jungles even in the dark of night"(Naik, 1980). According to legend, they have been blessed by Lord Shiva and have access to the jungle like a sher, meaning tiger or lion. The Saharia also call themselves, 'Bhil Thakur', which shows their consciousness for achieving higher status in the society. Parik (1975) writes that the Saharia regard the Bhils, Bhil-Bhilade tribes as the branch of the Saharia. Saharia tribal people consider Biju Bhil as their ancestor who came to settle here from the jungles of Aravali and Satpura many years ago and since then remained here. Saharia name has originated in the Moghul period. From the olden days this region was within the Moghul empire which is confirmed from the fact that the fort of Shahabad was built by Sher Shah Suri. Also Abul Fazal has praised Saharia for their faithfulness and industriousness in his book 'Ina-a-Akbari'. Here one question arises naturally that 'Bow and Arrow' are familiar arms and ammunition for the Bhils whereas Saharia possess axe. But this doubt should be removed considering the fact these people used to reside in thick forests. On seeing the animals and the region which has come up in the last few years it appears that in the olden time there would have been very thick forests and the number of forest animals also must have been very large. In this thick forest the use of bow and arrow does not seem to be so easy. Several such examples have been prevalent where young Saharia have killed tiger, bear in an encounter successfully. This shows that they have given up bow and arrow only after coming in the jungle. There is also another belief prevalent that Saharia have originated from 'Aryavarta' i.e. India. Sah-Arya means with Arya, i.e.Lord Rama, the famous king of Ayodhya and great man of epic literature, Ramayana. The Saharia believe that in Ayodhya they were influenced by Lord Rama and has surrendered themselves to him. Even in the absence of genuine historical account it may be stated that the Saharia have been one of the first settlers of Rajasthan. Tod (1839) has mentioned them along with Meenas, Bhils and Gujjars as the primitive dwellers of the region who "are still but little removed from savage life".

Distribution

The tribal (Saharia) belt in District Baran lies beyond the Parbati river in its eastern part. This part of the District is surrounded by Saharia occupied Districts of

Madhya Pradesh on three sides, to the north by Morena and Gwalior, to the east by Shivapuri and to the west by Guna which forms a major portion of the Saharia tract. The Saharia of Rajasthan are concentrated in two Tehsils of Shahabad and Kishanganj which also coincide with their Panchayat Samitis. The Shahabad and Kishanganj Tehsil can be divided into two distinct divisions (1) Upreti, a plateau or the uplands formed of the villages from Kelwada in the west to the Samarania, Mundiyar, Mamoni, Kaloni, Baint, Sahahpur etc. in the east upto Shahabad and (2) Taleti, formed of the low level villages beginning from Shahabad to Deori, Bamagaon, Faredua Taleti and Kasba Thana. The area is characterised by parallel drainage and all slopes are westward. Lohai, Andheri, Sukhan, Dubraj, Bilas, Bomi, Kul, Charoli are all ephemeral rivers. The Kunu is a perennial river. It follows from south to north and drains the Taleti part of the area into the river. The Kishanganj Tehsil forming the Western part of the 'Shahabad Panchayat Samiti' does have some portion of the Upreti but the rest of the area in the Tehsil is plain, with deep soil depth and higher water table. In Taleti, the altitude of the country sinks abruptly from some 1600 to 1000 feet, thus forming a trough to which the name Taleti or lowland has been given. Due to thick forests and hills this portion is not very fertile. It is mostly a poor jungle. In Upreti, villages are connected with pucca road whereas in Taleti the villages are not connected with pucca road. The villages are interiorly located in Taleti region. There is also an 1800 feet high hill in Shahabad known as "Mamuni ki Pahadi". The Upreti area is irrigated and more fertile than Taleti. But the climate of the eastern part of Upreti is very unhealthy and the water supply is less assured. Hence, this portion is less populated. People who live here lead a hard life. The western part of Upreti is quite fertile and well inhabited. It has a belt of yellow and black soil.

Physical Features of Saharia

Generally Saharia are broad chested and black skinned people. They have flat nose, curly hair, thick lips, narrow head.

Dialect

Saharia generally speak the language of the region. Locally the language is known as 'Khariboli' which is a mixture of Hindi with Rajasthani and Malwi accent. The 'Devanagri' script is used by them.

Clans

Saharia is an endogamous group divided into 56 exogamous patrilineal clans, each clan named after some ancestor or a place to which they originally belong. The following clans have been found among Saharia: -

Bedgor, Badralia, Bhororia, Bareliya, Bhilodiya, Chackrya, Chowdarya, Dacharia, Digoria, Dosania, Dediya, Devaria, Debariya, Dareliya, Garwar, Gairchina, Gogaya, Haleriya, Jarkoliya, Jaswariya, Jajwar, Khagania, Kuhar, Kanachfriya, Kheti, Khanwar, Kheliat, Kundawar, Khadeliya, Kareaniya, Khadiya, Karuriya, Kanwal, Kaluaj, Lagayahi, Mogriya, Navtriya, Navriya, Nimaria, Nolani, Parenatia, Pateja, Parondiya, Porandi, Rajouriya, Rewar, Raichohan, Ragpita Chohan, Rathwar, Sanoriji, Silwar, Somani, Sopariya, Solvia and Solanki. From above it appears that except Bhilodia which can be directly linked with Bhils, there are atleast four clans which appear to bear the names of Rajput clans like Rai Chauhan, Solanki, Devaria, Ragpita Chauhan.

Each clan is again sub-divided into a number of exogamous lineages. Most of the clans are totemistic in nature and worship the specific deity, object, plant, animal etc. Some clans are territorial in origin. The members of a clan known as 'gatia bhai' believe themselves to be descendents of a common ancestor. Usually a Saharia village is dominated by people of a single clan. The clan is patriarchal. The totemic objects are respected by the people who abstain from killing and eating these objects. All the clans are in same status except four clans (Rai Chauhan, Solanki, Devaria, Ragpita Chauhan) which enjoys a little higher status because they belong to clans of Rajput. A smaller unit is known as lineage i.e. 'kutum'. Members are genealogically related. They describe themselves to be a part of one house. A lineage is characterised by active cooperation among members during rites de passage and other socio-economic activities. The fission in a clan or lineage may be caused due to famine, drought, epidemic and migration. Now-a-days, the Saharia migrate frequently from core region to plain areas to pursuit their livelihood.

Hamlet Pattern

Saharia do not have disperse pattern of living as found among the Bhils of Southern Rajasthan. They do not have even separate and scattered hamlets. However, the hamlets are slightly far off from the main villages. Each village has a cluster of Saharia families in a separate hamlet known as 'Saharana'. In the 'Saharana' no other caste people live. In the centre of each 'Saharana' they construct 'Umbrella type' shelter house which is called 'Panchayat Bungalow' where all Saharia sit together to discuss, gossip and settle their disputes. Shoes are not allowed in the Panchayat Banglow. Women are also forbidden from entering the hall. It is believed that as women come from other villages, secrecy cannot be expected from them.

Flora

Some of the important trees, shrubs and grasses available in these forests forming the Flora include Basis latifolia (Mahua), Buchanania latifolia (Achar), Boswellia serrata (Guggal, Halar/Solar), Lennea coromondelica

(Gurja), Termenalia belerica (Bahera), Diospyros melanoxylon (Tendu), Terminalia arjuna (Kora), Albizzia proura (Sirish), Eugenia jumbolana (Jamun), Anogessius latifolia (Dhonbra), Acacia leurophbora (Khenjra), Salmalia malabarica (Sermal), Acacia nilotica (Babool), Ficus religiosa (Peepal), Ficus racemosu (Umar), Zizyphus jujuba (Bor), Butea monosperma (Kesuda/Khabro/Padash), Magnifera indica (Amba/Mango/Kairi), Cassia auriculata (Anwala), Carrisa spoinarun (Karonda), Asparagus adscendens (Dholi moosli), Schina neniosiam (Sen-senta), Azadirachta indica (Neem), Michalia champaca (Champa), Ocimum basilicum (Tulsi), Pratea frandosa (Dhak), Acacia catechu (Khair), Cassia fistula (Bahania, Dhedlya), Sterculia urens (Kendya, Kandol), Angoliosus pendula (Dhao, Dhuwali), Solanum nigrun (Makoy), Nyctanthes arbortristis (Harshingar), Lannea coromandelica (Gurjan), Aegla marnelos (Bel), Staphegyne parvifola (Kadam), Dalbergia sissos (Shisham), Terminalia sometosa (Sadaria), Syzajgium cumins (Jamuna), Salmalia malbarica (Semal), Arisid depresses (Lapla), Apluda mutica (Polard), Themeda quadrivulins (Ratarda), Heteropogon coatortus (Surwal), Dichanthium annulatum (Korar), Hrvsopogan fulous (Bhalbi).

Fauna

Among wild animals, mention must be made of Panther (Panthera pardus) Tiger (Panthera tigris), Chinkara, Chital (Cercus axis), Nilgid (Boselphus tragecamelus), Sambhar, Sloth Bear, Wild Board and Common Hare. In addition, Common Langur, Stripped Hyaena, Jackal, Indian Fox, Jungli Cat, Indian Porcupine, Common Mongoose, Hedge Dog, Stripped Squirrel, Bat and Bush Rat are also found in the forest area. Common birds include Baya, Koyal, Vulture, Parrot, Jungli Crow, Bulbul, House Sparrow, Owl, Kite, Green and Common Pigeon, Sand Grouse, Peafowl, Common Quail, Jungli Bush Quail,

Grey Partridge, Saras Crane, Common Sand Piper, Little Egret, Common Teal, Snips, Painted Partridge, Jungli Babblur, Black Strongo, Paradise Fly Catcher, Red Walted Lapuring, Magpie Robin, Indian Hyna and Hoopoe. The most common poisonous snakes are Cobra, Russel and Viper. It is also interesting to mention that a large variety of fish is seen in the rivers of the area. The major carps consist of Catla (Calta calta), Rohu (Labio rohita), Mrigal (Cirrhina marigal), Lobeo calbasu and Mahaseer (Barbustor), while Chal (Chelabuccala), Puthi (pantiusticto), Bangar (C.reba), Bala (L. pata) and Dera (L.dera), are the important varieties among the minor carp. Singhara (Mystus seenghala), Lanchi (Waollago atta) Magur (Ompoic bimaculatus), Siland (Silonia silondia), Gonch (Rita rita) Girraj (Chena marulius), Sawal (C. Puntatous), Kabra (Charmasiriatus), Sate (C. gachua) and Balm (Eel) are some of the kinds of Cat Fish available in this tribal region.

Marriage

Marriage is a socially sanctioned relationship between a man and a woman with some objectives, v.i.z., sexual cohabitation, procreation, economic cooperation and common residence. Saharia is an endogamous group but each clan in it is exogamous. They believe in totem and taboos and observe them during rituals and ceremonies of marriage. Marital alliance between the mates of same clan is prohibited because by virtue of their affiliation to a common totem. The members of the clan would be reckoned as brothers and sisters. Exogamy based on clan is strong among the Saharia. Members of different clans are free to marry one another. It is customary among the Saharia that one cannot marry within its own clan, one's mother, father's mother, mother's mother clan. So a person has to leave atleast four clans for marriage. Agnates are strictly prohibited to marry one another. Monogamy is the general rule of the society. Polygyny is rarely practiced only by the well-to-do families. Usually Saharia follow the marriage by negotiation. Besides, marriage by elopement and by exchange occur occasionally. Since no bride-price is required for such type of marriages, a nominal fine is to be given to the village council. Cross-cousin marriage is unknown to them. At present junior levirate is not practiced, however, junior sororate is allowed. Marriage between known lineage is strictly prohibited. The symbols of marriage are vermilion on the hair parting, thati and toe ring. Bride price is paid in cash and kind.

The tradition of getting married inside Saharia tribe has remained very rigid among them. Inter-tribe or Inter-caste marriage was a punishable crime upto a few years ago. If a Saharia women married a non-Saharia person, he was compelled to pay penalty to Saharia Panchayat and have to eat and drink with Saharia, otherwise the woman used to be out from the tribe. With changing times, the restrictions on getting married outside the tribe have also started reducing. However the cases where Saharia women got married to persons of other caste were rare. In the past child marriage was prevalent among the Saharia. Now Saharia do not favour early marriage particularly child marriage. The average age at which a girl is married varies between 15 to 18 years. For boys the age varies between 16 to 20 years.

Widow marriage or remarriage called 'Nata' is permissible among the Saharia. A widow can be married only to a widower or to a divorced person. Even an unmarried person can marry a widow and vice versa, but such a union is not much favoured by the society. When a widow gets remarried or a married woman falls in love with other man and wants to live with him, the new husband has to pay 'Jhagada' (an amount against marital expenses) to the previous husband or his family.

Family & Kinship

Family is a unilaterally related group consisting of a man, his wife and unmarried children. Sometimes, married brother or son-in-law, their wives with children or married son live together. Nuclear type of family is predominant though extended type also exists. They practice patrilocal as well as neolocal residence. The authority of a family lies with the eldest male member. He governs the socio-religious and economic cooperation. He is the chief agency of social control. Land and other properties such as ornaments, cash, utensils etc. are equally divided among the sons during the old age of father. A woman avoids her husband's father as well as all brothers of her husband. She does not talk directly to them and keeps veil. Joking relation is maintained with her husband's younger brother and sister. A man avoids his wife's mother and all sisters of his wife. He shares a joke with his wife's younger sister and brother. Mother's brother and father's sister have obligatory relations. Inter family relations are community based and reciprocal in nature.

Mode of Inheritance

Succession goes to the eldest son. The women have no right to inherit property. She plays an active role in domestic sphere and family management. She goes for earning and contribute to the family income.

LIFE CYCLE CEREMONIES (RITES DE PASSAGE)

Birth Rituals

Customs related to birth are the same for both male and female children. Mostly the sweepress conducts the delivery. Just after the birth of a child, a messenger is sent to girl's parents with sweets. On the third or fifth day after the delivery the mother and child are given a bath and taken out of the house for ritual purification. It is called 'bahar nikalana'. After delivery, the mother is given jaggery and mahua water to clean internal organs and impurities from the body if any. During the maternity period, the mother is fed on dalia prepared from jaggery and jawar or wheat floor. All oily foods, meat and spices are tabooed. The collective pollution involving family members is observed for three days during which contact with outsiders is forbidden after which the house is sacralised with cow dung. An elderly woman ties an amulet round the waist of the new child and conducts the simple name giving ceremony. Sweet balls are distributed. On this day the girl's parents bring clothes. They are served with meals. It is called 'pachlana'. The Saharia women take to household chores three weeks after the delivery. But if there is no other woman to take care of the house, the Saharia woman starts performing household chores immediately after this ceremony.

The cutting down of the hair for the first time has ritualistic importance in Indian societies, specially of a male child. This ritual is known as 'mundan sanskar'. The Saharia also perform it for male as well as female children. A barbar is called for this purpose who uses a pair of scissors for it. He charges Rs.20/- to 25/- depending upon the economic condition of the family.

Marriage Rituals

1. **Sagai or Lagan:** In Saharia, the initiative for setting a marriage is generally taken by parents of the bridegroom. Decisions regarding selection of the partner are always taken by the parents. There is no system of looking the bride before marriage. The performance of betrothal is very simple. In two donas (leaf bowls), they put grass, rice and turmeric and a copper coin. Then the 'Panch' touch the bowls and give their blessings.

After this the bridegroom's sister-in-law and Patel (headman of the village) exchange the bowls seven times and then each go to their village with the bowl in their hands. The guests are farewelled by putting turmeric 'Tilak' on their forehead. The girl's father presents a turban to the bridegroom's father or elder brother and greets him by saying 'Ram Ram'. The bowl is taken inside the house with songs. Untill the arrangement for music is made, the bowl is placed in a house in the neighbourhood. This is called 'Lagan'. The day of marriage is decided by Panchas at the time of engagement without consulting the priest.

2. **Bindayak:** On the day of 'bindayak' the boy applies on his body a paste prepared by mixing turmeric, floor, oil and water. His aunt and sister-in-law put mehandi (henna) on his hands. Then the bridegroom worships Ganeshji. Afterwards the aunt and sister-in-law tie 'moli' thread on his right wrist for which he pays some money in the form of 'neg' to them. From this day singing of songs starts which continues till the day of procession. Everyday the bride-groom worships Ganeshji.

3. **Tel Bithana:** The bridegroom is seated on a wooden slab on the day of 'tel bithana'. Then both aunt and sister-in-law touch his body, beginning from feet to head with hands dipped in oil. This is repeated seven times. The custom is again repeated on the day prior to marriage but the direction of applying oil is reverse, i.e., from head to feet. This is called 'tel utarana'.

4. **Mandap:** A 'mandap' is built in the girl's house where the actual marriage takes place. Four poles are driven in the ground and bamboo staves are

placed diagonally on them. These are covered with leaves and twigs to form a shed. The four poles are from Jamun trees. On both sides of the entrance of the house, four pitchers, one above the other, are placed. These are tied with red cloth or `moli' thread. It is called chavari bandhana'. On this day, the bridegroom's father gives a feast to the whole community in which jawar' on `bajra ki roti' and curry prepared from the same floor are served. After the feast the procession (`barat') starts for the house. Women and children also join it.

5. **Padat:** The 'barat' reaches the bride's house on the day prior to the marriage. This day is called 'Padat Ka Din'. The bridegroom goes to bride's house in a 'palaki' (palanquin) if one can afford it. Use of a bullock-cart is more comon for this purpose. During the procession, instead of employing a band people beat 'thalis' (copper plates) which makes a sing-song noise. But now, use of a band to some extent has also started amongst the Saharia. Most of their customs have a resemblance to Hindus. However, a significant deviation from Hindus can be seen during the ceremonial touching of the toran. The Saharia bridegroom is not mounted on a horse-back but is simply lifted up by his Sawasa, i.e. sister's husband.

6. **Phere Padna:** For performance of 'pheras' (to take rounds) neither a priest nor a nuptial fire is lightened. This ritual has important differences from Hindus amongst whom presence of a priest and nuptial fire are important for 'pheras'. The bride's sister-in-law ties the 'gatha-joda' of the bride and bridegroom and brings them in the 'mandap'. The bride sits near a pole of 'mandap' and bridegroom makes six circles round the pole.

The seventh circle is taken by the bride. The Patel of the village assists in the process of promises taken both by bride and bridegroom and this completes a Saharia marriage. Then, the bride and bridegroom go to 'Jawasa' (place where the barat stays) where the bridegroom's father gives his blessings to them. After this, the bride and the bridegroom worship their god and goddess and seek their blessings.

7. **Barat Vidai:** When the marriage party departs on the second day, a tilak is put on the forehead of each of its members.

8. **Gauna:** After marriage the bride stays in the in-laws house for about a week and then comes back. She again goes to her in-laws' house after one year. At this time, a ceremony is performed which is called 'gona'. A large number of guests from both sides gather at the bride's place for the farewell of the bride. Women of the bride's side sing vulgar songs and make jokes. Among Saharia some token money is given for each song known as 'Geet Thali' which is given by the bridegroom to the ladies who sing. The bride puts on lehnga and chunri and is given farewell along with the bridegroom. The bride-groom pays 'neg' to the women for their songs. The bride wears new clothes at the time of her farewell. Vermilion is used as a symbol of marriage by the women. Residence after marriage is patrilocal.

Death Rituals

The Saharia cremate their dead. On the third day also, a feast is arranged for relatives only. Either on the third day's feast or on thirteenth day's feast all male family members of the deceased have to cut off their hair. For thirteen days the close relatives of the deceased stay and eat at his house. On thirteenth day, a feast is given to the whole community (of same clan). Offering hair to the soul of deceased is an important ritual connected with death ceremonies among most of the societies in India. The ashes of the deceased are put into Kapildhara (a pious river of the Saharia tract) or Sitabari (a religious place in Kelwara which has water ponds of historical origin).

Nata Custom

Divorce is permitted on grounds of adultery, maladjustment and insanity. The remarriage of divorced person is permitted. 'Nata' is very common among the Saharia and they do not consider it bad. If an unmarried person develops `nata' is (relation with opposite sex) then he is given a severe punishment. He is asked to give manda feast and on the same day he is asked to perform rounds around the female ear ring or a pitcher. Infact the women Saharia along with whom the man elopes is brought and her ear ring or a pitcher is kept under the mandap and the Saharia man is required to go round the ear ring or pitcher seven times. The system of taking seven rounds around the ear ring or pitcher is considered as atonement for sin committed by the unmarried man.

When a women deserts her husband and goes along with some other man even then clan is kept in mind, such cases are resolved by Punch Panchayat (five member committee) and Punch of all the villages participate in it. Both the man and women appear before Punch Punchayat. Punchayat awards them physical as well as financial punishment. Previously the punishment given was physical as well as financial punishment. Previously the punishment given was so severe that the women was beaten with shoes and forced to urinate. A small stone was kept on the head and over that a big and heavy stone was kept. But now the physical punishment has been reduced.

The fine imposed on the women is given to the previous husband. If the woman wants to return to the previous husband then both are punished i.e. the person with whom the woman has eloped and the previous husband to whom she has returned and who will not keep her. A false allegation of 'Pattal bhelan' is put on the married man. To ensure that the man who had eloped with the woman will not repeat the act, brother- sister relation is established between the two, before the Panchayat he touches the feet of the woman and offers her clothes to wear. In case he repeats the offence then the punishment awarded is twelve times more.

Social Unity

Social unit among the Saharia is based on primitive tribal locations and the social unity among the other communities are based on their belonging to higher and lower castes. In this regard Saharia do not consider themselves to be an origin or a part of the Muslim or local Hindu community. On account of this reason the importance of social unity among the Saharia gets increased. Majority of Saharia usually like to have small families. In general the custom prevalent in Saharia is that after the marriage the son is given a separate hut and either he is given a share in land to earn his livelihood or both husband and wife work as a labour to earn a livelihood. According to Hindu norms a joint family is basically considered an ideal family in which the eldest son has got a special responsibility. As far as the responsibility of the eldest son is concerned the same norm is followed in Saharia community. After father's death the eldest son performs the last rites. The eldest son gets separated just after marriage. While distributing share of property one part of the share of property is kept separately for the parents. Youngest son after his father's death continues to live in the same hut. It is, therefore, natural that parents share of the property as also the parental house are acquired by the youngest son. Since he gets major share of the property it is considered his responsibility to marry his unmarried sisters and brothers. It is also expected from other brothers that they should share such sort of responsibilities. It is considered their good will if they extend a helping hand but it is not considered as their duty.

Village Unity and Habitation

The Saharia people in every village live in a separate hamlet. Their dwelling is known as 'Thok' or 'Saharana.' For every village there is separate 'Saharana'. In the middle of 'Saharana' there is a house which resembles the shape of an umbrella which is called 'Panchayati Bangla'. Around this Bangla huts made of grass, wood bricks etc. are constructed in which Saharia families live. All Saharia assemble at this Bangala and entertain themselves. They also resolve their disputes at this Bangala. In the assembly the influential man of the village is usually the head who is called 'Patel'. To carry shoes in the Bangla is banned. Women folk are also prohibited to go to these Bangla. This is so because it is believed that women come to this village from outside (i.e. another village) after marriage and they are, therefore, not considered member of this village. Patel's appointment is based on heredity criteria. But if he is not fit or suitable or found guilty in a crime then he can be removed by the influential people of the village who can then appoint another Patel.

Confrontation and Isolation

The Saharia area was a jungli nizamat under the erstwhile ruler of Kota State. Jaswant Singh, erstwhile ruler of the State, with a view to develop the revenue of the area, allowed Sikhs and Jats to settle in this region. The land was given to these immigrants on nominal rates.

The immigrants occupied the irrigated area of Kishanganj tehsil and pushed Saharia deeper into the forest, and dislocated them from their lands. In fact, Saharia have been very much exploited by these powerful groups to whom they were forced to mortgage their lands.

Economic Disequilibrium

The Saharia maintained ecological equilibrium with their environ for ages, despite their low level of technology and axe cultivation. They could sustain themselves as gatherers of minor-forest produce and agricultural labourers. With the emergence of stronger groups, who invaded the region in search of better lands, the ecological balance was disrupted and forced Saharia to subsist on poor and scanty natural resources. Over-exploitation of the rich resource Saharia by the outsiders, made Saharia poorer. Saharia struggle for existence is going on. More soundly constituted groups like the Sikhs, Jats, Kirad, Rajputs drove away Saharia to the areas less favourable to human living. They have impoverished the Saharia.

Literacy

The Saharia have a very low literacy rate. They have achieved only 4.72 per cent literacy (1981 census), the female literacy rate being only 0.72 per cent, and of the males 8.60 per cent. In the rural areas they now send their sons to schools but mostly they drop out. So also the case with girl students.

Drinking and Smoking Habits

People of Saharia tribe are a victim of exploitation and illiteracy. They suffer from various types of diseases and for their bread and butter depend on others mercy. Inspite of being victims of poverty and malnutrition they do hard labour to earn their livelihood. At the same time they are complacent as well, and search for a work to earn livelihood only when the hunger stares them on their face. Because of extreme financial hardship and poverty the Saharia tribal people of Shahabad region have developed a habit of drinking which is resulting not only in reducing their labour output capacity but also making them more vulnerable to various diseases. Drinking wine has been a custom with Saharia and these people prepare local wine with the help of flowers of 'Mahua' in their households and consume it. And they also sell this self prepared wine on cheap rates to earn a livelihood. Even in today's date these Saharia prepare wine in their houses and drink it. Other members of the household do not hesitate in taking wine casually or on festivals. In Shahabad region not only tribal (Saharia men) but their women folk also drink wine. Their houses may not have grains of food but drinking wine has become a mental obsession with them. Men drink to relieve themselves of their daily tiredness while women drink to relieve themselves from their suffering. After or along with drinks they usually take simple roti (bread), chilly or chatni (sauce) before they go to sleep (or retire to bed). They are also fond of smoking in addition to drinking. These tribal people quander a major part of their earning in drinking and smoking bidis. They have nothing to take from or give to the world. Since these people live an independent life and take to earning wages from the young age which results in picking up these habits at the young age from their friends and colleagues. At present smoking has become very common in young boys and girls of this tribe and in the last few years 'Gutkas' have also become familiar with young men and women. Although most of the tribal girls who work at various places take lime mixed with tobacco. Approximately 90 per cent tribal men take wine. 70 per cent of women are also victim of one or the other type of addiction. This had also led to increase of tuberculosis considerably. By nature these tribals are innocent. Because of their long continuous exploitation over the years as also injustice meted out to them because of their being deprived of forest rights by the Govt. and also because these tribal people have no other opportunity to work to help them earn a bread, they have been driven to become slaves of drug abuse and drinking, adversely affecting their physical and mental development. Though the state Govt. has started 'Saharia Vikas Pariyojna' for the upliftment of the poor and down trodden Saharia tribe and have spent crores of rupees on pariyojna scheme yet their poverty has not been eliminated and they have not got rid of their habit of drinking either.

Economy

Ecology plays an important role in shaping the mode of livelihood of a people. The Saharia prefer to live nearer to forest and depend considerably on forest products. The impact of ecology on their economy is indispensable and the environment has significantly influenced on the economy of the Saharia. In the past, the Saharia were nomadic in nature and used to practice shifting cultivation, food gathering and hunting. The present day's Saharia live in the stage of food producing economy. They are mostly settled agriculturists and supplemented by domestication of animals, fishing, hunting and collection of forest products. Moreover, many of them have become wage earners. A trend of occupational mobility from traditional to modern is observed among the Saharia. They have shifted from clan oriented (bond type) food gathering stage to pre-agricultural stage to family based settled agriculture to a dependent economy of wage-earnings. The production of the 'kharif' crop mainly depends on rains and not on irrigation. The 'kharif' crop constitutes maize, millets like 'bajara', 'rali', 'jowar'. The 'rabi' crop constitutes wheat, barley, gram, mustard, coriander, onion etc. Land and forest are the main natural resource for Saharia. Traditionally they were hunters and gatherers. Now they primarily live on collecting forest products and farming. But their access to forest is limited and the small plots of land which they possess are not viable for their purpose. So they have to take daily wage labour in agrarian as well as in other sectors. Secondly, they also migrate to urban industrial areas in search of employment. Efforts are being made by the governmental agencies to rehabilitate them. They are also given jobs in lean seasons. The Saharia do not like to work as 'hali', i.e. contract labourers, because they get low wages and ill-treatment from the employer. Sometimes due to some unavoidable circumstances such as rites de passage or illness, the poor Saharia work as contract labourers for a specific period. In the past, the men were engaged as bonded labourers in the house of land-owner and the remuneration was very low. Usually men and women are engaged on 'rajyama', i.e. daily wage basis. In the past, Saharia used to get free meals, clothes and nominal payment. But now the practice of bonded labour ('hali') is no more visible in tehsil or block areas but it is still focussed in the interior areas in modified form. Method of cultivation by the Saharia is not much different from that of the neighbouring people. Mainly winter crops are produced by the Saharia. Share-cropping is not popular among the Saharia, because most of the Saharia are poor. They do not have bullock and agricultural implements. Most of the women work as daily wage earners or sell firewood in the market. Usually they get a sum of Rs.10/- to Rs.15/- from a headload of firewood.

Every year a good number of Saharia migrate temporarily to neighbouring wheat producing area at the time of harvesting season. Seasonal migration is a common feature among the Saharia. They harvest on contract basis and get the wage in kind (wheat) instead of cash. This brings them a good collection of wheat grains, which last for three to four months for family consumption. Almost all of them were eager to earn cash by selling forest products during the lean season. Sometimes, they are also engaged as labourers in forests and for road construction. Indebtedness is common among the poor Saharia. The factors generally responsible for poverty are the indebtedness, the low income, landlessness, lack of education, habit of drinking, frequent remarriage, huge expenditure on rites de passage and in other ceremonies. Such deficit of money are met by borrowing money from money-lenders or by selling property. In the past the economy of Saharia was forest-based. The collection of forest produce and hunting were their main occupations.

Earlier agriculture was of secondary importance and was practiced by few Saharia only. During state times, they were forced to work as bonded labourer by both Zagirdars and rulers. No Saharia could ever think of escaping it. They were forced to do all types of labour in the form of beggar. This included work's in the fields, cutting of fuelwood etc. For their subsistence, the Saharia were provided with one kilogram wheat or any other grain every day. After independence due to various measures taken by the government for the upliftment of Saharia, their economy has undergone considerable change. They were provided with land and other subsidies. Now labour and agriculture are their main occupations. Not only in the past but even now the forests are an important additional source of income of Saharia. Saharia are engaged in various forestry development works and forest operations concerning of timber, firewood and other minor forest produce like 'Tendu Patta' (Diospyros melanoxylon), 'Bidi' Leaves, Gum, Honey and fruit of Amla, Achar (Bachanania latifolia), Mahua (Basis latifolia) etc. also forms an important source of income for livelihood. They are financially dependent for their livelihood on means such as cutting tree,wood in jungle, collecting gum from jungle trees, collecting honey, pickle, grass, 'Tendu Patta' and selling them at marketing centres. Now they have started working as agricultural labour.

The women of Saharia community contribute to the family income by working for wages. They also carry out household chores and play a significant role in family management. On one hand where these tribal women not only do their daily household work but on the other hand they have also to toil hard to earn their livelihood. After a very hard daily labour to earn livelihood these women get only jowar, maize or wheat, chapati and be content with it. Because of malnutrition their new born children suffer from anaemia, tuberculosis and other eye diseases. These children also meet an early and untimely end because of lack of or non availability of proper medical facility. Because of extreme poverty people of the Saharia community have no alternative but to work day and night as labourers in different and difficult environment and conditions such as road construction, earth crusher, cement concrete construction in brick kilns, machines for cutting wood, bridge and building construction to earn livelihood for their children. These families spent their summer and winter nights on the footpaths or vacant verandah or shops but have to search for a tine top shed during rainy season. In the case wife and children are also accompanying him then he has to prepare a shed (shed made of stone-mud walls with roof made of cut wooden logs covered with rin sheets). They keep their bread in cloth bundle they keep with them while they work and look at the passers by with hope and expectation. Saharia women and men folk say that they do not know while leaving their house in the morning where from they will earn their bread that day. They are content with their wages which get increased with the increase in prices.

SAHARIA LABOUR HARD TO PLUCK TENDU LEAVES BUT GET HALF THE WAGES

In Kishanganj-Shahabad tehsil of Baran District the work of collecting tendu leaves gains momentum at a high pitch in the month of May-June. Groups of children, women and aged men reach the jungle early in the morning to collect tendu leaves. After getting free from the cultivation work the families of farmers and labourers collect tendu leaves to earn an additional source of income. On hearing the news of tendu leaves collection the working atmosphere is created in the village and a family earns nearly three to four thousand rupees during this period of leaves collection. Elderly people who are unable to walk also get busy in making bundles of these leaves. After plucking these leaves, bundles of fifty leaves each are prepared and taken to an appointed place by the contractor in the evening, where business men give them wages for their labour and settle the account. Every one is aware of the commercial importance of fruits and also the wood that is being used as a building material and is obtained from these trees. There are other trees whose leaves are also useful in industry. One such is that of tendu also known as 'Timru' in local language. In Shahabad and Chhabra region of Baran district leaves of tendu patta tree have a special importance keeping in view their usefulness. 'The botanical name of this tree is 'Diospyros melanoxylon', which is a member of 'Abnacy' family. The leaves of this medium size tree are green. In jungles this tree is promptly recognised because of the protruding figure on the back of the trunk of this tree which is very much similar in shape to that observed on the back of a crocodile. Leaves of 'tendu patta' trees are plucked in the beginning or middle of May and this activity goes upto middle of June. In the summer season green leaves are collected and dried. Even after that the leaves become flexible because of moisture. An officer of the forest department is of the opinion that villagers are less aware about the usefulness of tendu patta leaves. That is why sometimes such leaves are found in bundles which are not helpful in making a good 'Bidi'. It takes about one and a quarter month for the leaves to grow on a Timru tree. If the leaves are red in colour they become black on drying. Such leaves are broken while preparing a bidi. If the leaves do not get broken they become hard, an effort is required to smoke such bidis, though there is no difference in tests. Two bidis can be prepared out of one leaf. A young forest officer is of the opinion that to get good revenue out of this tree it is safely and carefully preserved. Tribal people carefully do the job of plucking maximum leaves. After plucking green leaves, bundles are prepared out of these. Then these bundles are dried for about eight days. During this period bundles are turned up side down every day so that leaves get dried properly. After the leaves get dried bundles are prepared out of these leaves and they are wetted in water. Wet bundles do not get broken when put while filling in bags. These are kept in the sun for eight days and after that the leaves become suitable for making 'bidis'. For making the bundles dry they are set in a particular fashion. Dry bundles are filled in bags. A bag containing hundred such bundles is known as 'Manak' sack. These are taken from the shop to the godown under supervision of forest officer which is posted there for the purpose. According to information available with the forest department some contractors get 'tendu patta' leaves collected stealthily from Government centres. These contractors deliberately manipulate that leaves collection of four to five centres are not given on contract so that tendu leaves could be got collected stealthily from these centres. According to available information it is reported that every year tendu leaves are stolen from the forests of Shergarh. This forest falls within the boundary of Amlavada to Ataru tehsil. Its area is 49 sq. kms. Contractors keep an eye on the 'tendu patta' leaves in this region of 'Abhyarnya' because the quality leaves are found here. Bidis made of these tendu leaves are purchased even by business men. It has been intimated by the sources that forest department loses revenue because of tendu leaves being plucked in this region and law is also broken. Daily routine of forest animals is disturbed in 'Abhyarnya'. Panther, pig, deer, tiger and other forest animals get disturbed during plucking season. Tribal people have been able to earn bread and butter because of this tree. Tribal people give as much regard to this tree as they give to 'Mahua' tree. At the time of drought in jungles, in difficult terrain, when other relief measures are not available, at that time, the job of collection of tendu leaves proves to be a boon to the people, from the nature. It is such a profession where labourers were getting daily wages in the previous years but this year (2000) contractors have given jobs on a weekly basis so that wages can be distributed at the end of every week. Daily labourers are on the verge of dying of hunger. Because of the neglect of forest department tendu leaves are not able to reach the shops and middlemen with the connivance of producers of 'bidis' get their godowns filled up with dry leaves during the night. All the workers of the forest are aware of this fact. A usual exploitation of labour is continuing today uninterrupted. Though minister of forest and the staff and officers under his command have not been able to remove irregularities it is felt that government should give proper facilities to tribal people who are engaged in this profession. There is also need that government should give adequate attention and incentive to accentuate this profession. A lady labourer connected with this profession said that 'saheb' does not pay wages in cash but hands over a chit which if it is lost amounts to throwing water on their entire labour but they pleaded helplessness saying that they have to perforce content with what ther are getting. Girdhari Lal of Harinagar village is of the opinion that in tendu patta profession contractors are exploiting the labourers financially. Contractors are exploitating the labour employed at shops in tendu leaves profession. It is a common knowledge that they (contractors) are collecting five bundles per bag extra in the name of forest officers. Labourers are suffering all the time and from all angles but district administration has not been able to put a full stop to the exploitation of labour. It is a point to be noted that in Nahargarh forest range four units of tendu leaves are operative and which are controlling fifteen shops, where 15,000 labourers are getting benefit from this profession. Labourers on the shops have told that contractors/agents forcefully take additional bundles. At the same time persons engaged in counting also count the bundles less in number. It is also said that they treat/ handle the labour roughly. Contractor purchase the bundle at the rate of Rs.30/- per 100 bundles. It is learnt that contractors on the borders of Madhya Pradesh (MP) purchase tendu leaves in black market. Till date no ban has been put on the 'tendu patta' leaves that are brought from across Madhya Pradesh. Sources confide that tendu leaf contractors take into confidence the police and forest workers posted at the MP borders and successfully do black marketing of these leaves. Labourers say that contractors/agents who count the bundles, return the bundles by saying that these pertain to night period, so that these could be counted at half the wages.

It is for consideration that forest department has a large parafernalia to prevent such irregularities but in vain. One official of the forest department admits that corruption has continued in the past, is still continuing and will continue in future as well. It cannot be stopped and that they are helpless.

TRIBAL SAHARIA WILL NOW BE ABLE TO COLLECT GUM FROM THE JUNGLE

State tribal Saharia used to roam from one Jungle to another in search of gum tree in order to make both ends meet. They will now be able to have a sigh of relief as they will be able to collect the gum. For the last two years state Govt. had put a ban on the carrying of gum (by plucking from a tree) in the tribal Saharia region of Shahabad of District Baran. This ban has now been lifted by the present Chief Minister. It is important to note that main occupation of Saharia settled in Shahabad and Kishanganj region of Baran district is to collect gum from the jungle trees and work as a labour. That is how these Saharia have been earning their livelihood. But on Jan 5, 1997 forest department had put a ban on the collection of gum from the jungle. Since then Saharia were not able to collect gum. It is said that in the last two years several cases of collection of gum and its selling in a clandestine manner were also reported and forest department and tribal Saharia corporation also threatened these smugglers and since then this profession which serves as. means of the poor to earn a livelihood had been under crisis. District collector was of the opinion that after the ban, the black marketing of gum had increased. In such a situation he wrote to chief forest officer and apprised him of the complete situation but the forest department considering it to be a danger for the jungles did not agree to the proposal of lifting of the ban. When this matter reached the secretary to the Chief Minister and he was apprised to the facts who in turn explained it to the Chief Minsiter who took it seriously and after that the orders to lift the ban on plucking the gum from 'Khair', 'Dholera' trees of Shahabad region of the district were issued. It is said that after lifting the ban on collecting and selling of gum from the jungle trees the Saharia tribe of Baran district has received considerable economic relief.

Material Culture

The houses of Saharia usually contain instruments of agriculture, beauty aids (items of make up), food items, utensils of brass and copper, hearth, chakki (hand grinder), broom, earthen and iron cooking plate, saw-axe, wooden basket, lantern, pitcher, spade (axe), fish trap, Dholak, Manjira etc.

Political Organisation

The Saharia have a three tier panchayat system to settle intra-community disputes. There is a panchayat at the village level, in addition to the 'akadasia panchayat' for eleven villages and a 'chaurasia panchayat' for the whole community. Saharia panchayat is concerned generally with the control and regulation of relations between different members of society and to maintain tribal traditions, administration and rules. It can use physical force, punish a person and can impose fines. It consists of a headman or Patel who is assisted by a council of elders. The post is hereditary and the succession follows the male primogeniture rule. The number of members depends on the population of the village. It consists of five members (one Patel, two Pradhan's of I and II rank, two Barai of I and II rank). The headman or Patel as locally known, takes decision to call a meeting. He conveys his decision to Pradhan who passes on it to Barai. It is the duty of Barai to inform whole community about the place and time of the meeting. The Panchayat is reached to decision either through consensus or majority verdict and headman has no power to over rule the majority verdict. The headmen of many villages join themselves into a group which constitutes the main panchayat body or council. On major disputes all headmen collect to take the decision. The Saharia panchayat is very effective in regulating their socio-cultural aspects. All of these panchayats settle disputes and punish the offender at their respective levels. For minor offence cash fine is imposed and for a severe offence the guilty is excommunicated. The statutory panchayat looks after various schemes aided by State Departments.

Festivals and Worshipping of Gods

Besides following Animism (belief in spiritual being), the Saharia also profess Hinduism and with the same devotion they worship, Ram and Hanuman and have high respect for "Ramayana". They are true devotee of goddess Kali and during 'Navratras' they worship Bijasan Mata, Amba Mata, Balaji, Naharsingh and Bheruji and Tejaji as deities. They have faith in supernatural powers. The rituals observed by them are akin to the Hindu rituals. Saharia generally worship 'Tejaji'. They believe that if they worship 'Tejaji' snake does not bite, if a snake bites a person then they put a 'Bandh' in the name of 'Tejaji' which is opened on 'Teja Dashmi'. At that time poison spreads and the person on whom 'Tejaji' comes ('Janteyar') sucks the poison with his mouth and cures the person bitten by the snake. 'Jin' is also worshipped. This worship saves them from the curse of diseases. Apart from this, they also worship lord 'Hanuman', 'Kali', 'Ganesh', Goddess 'Durga' etc. Saharia celebrate the following festivals in general:

(a) Makar Sankranti

(b) Amavasya of Sawan

(c) Janmashtmi (Birth day of lord Krishna)

(d) Raksha Bandhan

(e) Holi

(f) Diwali

(g) Teja Dashmi

On the Diwali day cows are worshipped. Mehandi and colours are used to decorate the animals in an artistic manner. They are decorated with garlands of flowers. Following names are worth mentioning which are prominently used for decoration:

(a) Ghungermal

(b) Moria

(c) Karliya

d) Sansat

(e) Morga

(f) **Ghunger mal**: is a garland of leather which is put on the neck.

(g) **Moriya:** is tied on the face, it is decorated with koris.

(h) **Karliya:** is a necklace made of koris.

(i) **Sansat:** is called a rope made of feathers of peacock.

(j) **Morga**: is made of peacock feathers which is tied on the head.

In addition to this Saharia also worship 'Lakshmi'. The zeal Saharia have in celebrating the festival of Holi is not observed for other festivals. On the Amavas (no moon) day, stick ('Danda') of half of the month of Phagun is fixed on earth and for fifteen days songs of 'Phag' are sung. They dance and play. On the 'Dhulandi' day the youths play colour with the wives and brother's wives ('Bhabhi'). During these fifteen days liquor and meat are used in plenty. Jokes or teasing on these occasions is not considered bad i.e. considered civilized but if a person from other community cuts jokes or indulges in teasing women then quarrel starts. In addition to above festivals, Saharia also participate in fairs (melas) with enthusiasm where articles of daily usage are purchased. Such occasions also provide an opportunity to select life partners. Such incidences are commonly observed on these occasion in fairs (melas). Usually, in melas the boy and girl elope after taking fancy on each other. If a young women on getting an opportunity takes the first initiative and throws a bidi or any other article in front of another man or accepts and insists on him then Saharia man elopes with that women. Such initiative from a woman is considered as a challenge to the manhood of Saharia. This is the reason that some times married Saharia man deserts his wife and children to their own fate and goes away along with other woman. The mela of Kapildhara and Sitabari are famous among the fairs (melas).

SITABARI FAIR

In the tribal region of Sitabari of District Baran in Rajasthan annual fair 'Sitabari' takes plaes in the month of Vaisakha Sudi fifteen to Jyaistha Badi fifteen. Tribal Saharia grandeur of rich culture is clearly visible in the festival of Sitabari. Sitabari is a religious place where people earn 'Punyas' by having a bath in 'Lakshman kund', 'Lavkush kund' and 'Surya Kund'. Members belonging to all communities participate in the fair which has an average congregation of 50,000 people. Here a bathing on the eve of 'Somavati' Amavasya' or 'Shivaratri' has a special significance.

Various fairs ('melas') like Mataji, Phalodi Mata, Kawaleshwar Mahadev, Gopreshwar Mahadev, Shivaratri, Dusshera, Dudhiya Kheri Mata, Nahan, Dole, Akshaya, Balaji, Ramnavami and Tejaji are held in the Saharia region and Saharia participate in all these fairs.

Dress and Ornaments

Saharia men wear dhoti upto ankle and put on a white turban. In the past, majority of Saharia used to live with bare minimum clothes but now gradually they have started putting on shirt, half sleeve or sleeveless coat called 'Saluka' and keep long hair. Now gradually this custom is getting obsolete, youth of the new generation have started keeping hair in English style. Previously they used to be barefooted but now they have started wearing shoes. Women have started wearing multi coloured dresses like 'Ghaghra' and 'Chundahari' and 'Saluka' is there main attire. Ladies are fond of tattooing their bodies.

Women Ornaments

Saharia women and young girls are very much found of wearing ornaments. They decorate themselves with different kinds of jewellery and enhance their physical beauty. According to their economic condition they put in jewellery made of different types of metals such as silver, brass, gillet, raga, kansi and Jasta. They put on 'Bulak' on forehead, 'hasli' and also chain made of 'moonga' in the neck, silver or brass tokni in the ears and in the hands and feet they wear 'Gram Khajura', Kadula, Karghani, Jhalara, Bangle made of lakh in the wrists. These are the main ornaments of the women folk.

Man Ornaments

Men usually wear few ornaments, depending upon their economic condition. They put on ear rings or 'Murki' made of silver, 'Tabij' or 'Kanthi' made of white or black thread in the neck, or chain made of brass. They wear shirt or 'Saluka'. A person who is economically sound uses silver buttons in his shirt, although there are several people even today who use buttons made of leather.

Tattooing

In order to protect themselves Saharia men and women have tattooed their bodies. It is their belief that this act will guard them against diseases. Women consider this as jewels on their body. Saharia women folk like to decorate themselves with lipstic, ruse, cream, powder, kajal, bindi. They like to dress up their faces and hair.

Folk Tales and Songs

Saharia have a rich tradition of folk tales and folk

songs. They also have specific dance forms. On festivals and rituals they sing and dance. They also have decorative art like floor and well painting. Small drums and twin flutes are used by them during songs and dances.

SAHARIA FROM OTHER REGIONS

(a) Saharia from Bundelkhand (U.P.): The Saharia are chiefly found in the Bundelkhand and its adjoining areas. It is generally conjectured that the name is connected with that of the Savaras, the name in Sanskrit literature for the aggregate of the tribal communities like in the highlands of Central India and included the race like the Kols, Munda, Kuruks, Bhils, Bhuiyas etc. who live in the highlands of central India. They are also known in Bundelkhand region by another name i.e. Rawat, which comes from the Sanskrit word 'Raja-Duta' (King's messenger) or 'Raja-Putra' (King's Son). The Saharia of Lalitpur are divided into a number of exogamous sections (clans); such as Siransiya, Kodoriya, Thegotiya, Sanaurhiya etc. Some of these sections are totemistic in nature. The principal occupation of the Saharia is to collect fuel wood and grass from the forest and sell the same. They also work as farm servants. They chiefly worship goddess Bhawani but also have a number of local deities like Gonr, Margingha, Sanswar, Goraiya, Katiya, Thotiya, Somiya and Ahay Pal, many of which appear to be deified worthies of the tribe. In Uttar Pradesh, the Saharia have almost changed to a caste and completely lost tribal characters. In 1931 Census Report of Uttar Pradesh, Turner stated that "the Saharia of today must be classed as a low Hindu Caste, who have like Ahirs, Chamars, Gadariyas, Bhangis etc., their own

traditional occupation, their own place in the vast web of the Hindu caste system to which they have fully adopted themselves.

(b) Saharia of Madhya Pradesh: The Saharia of Madhya Pradesh are distributed in the Districts of Bhopal, Vidisha, Shivpuri, Guna, Datia and Gwalior. In this state they are notified as Sahariya, Saharia, Seharia, Seharia, Sosia, Sor and they number 261,821 according to the 1981 census. Of their total population, 6998 persons have returned from urban areas. Hindi is spoken and the Devanagari script is used by them for both inter-group and intra-group communication. There are fourteen patrilineal exogamous clans (Kuanwa, Karwar, Chouhan, Semrewa, Jirmeria, Kodlia, Kharewadia, Tataiya, Porai, Brishrewani, Deshbaria, Gidrania, Gorewa and Bansoria) among them. Crooke (1896) recorded thirteen gotra (Sirausiya, Kodoriya, Thegotia, Soanauliya, Rajauriya, Jachoria, Kusmonua, Sarosawa, Chakaroya, Chirauncha, Karwariya, Bagauliya and Sanhauriya) groups among them. Agricultural labour is their main occupation. Collecting firewood from the jungle and selling it is their subsidiary occupation. According to the 1981 census, 39.31 percent of their total population have returned as workers (55.72 per cent males and 22.08 per cent females). Of them, 49.95 per cent are agricultural labourers and 39.21 per cent are cultivators, 3.16 per cent are engaged in the collection and selling of firewood, 2.38 per cent are in mining and quarrying and the remaining 5.31 per cent are

engaged in other occupations, such as construction, household and other-than-household industries, trade and commerce. They are not enthusiastic about formal education for their children, and only 2.74 per cent of them are literate, the rate for the males being 4.99 per cent and for the females 0.39 per cent. The 1981 census records 99.89 per cent of the Saharia as followers of Hinduism, while 0.03 per cent belong to various other religions such as Buddhism, Sikhism. The followers of 'other religions' are 0.06 per cent, and 0.02 per cent have not stated their religion. Development measures have helped, to some extent, in alleviating poverty. The Saharia of Madhya Pradesh are different from Saharia of Rajasthan and Uttar Pradesh.

5

Health Status and Practices

Overview of Baran District

District BARAN, the place is situated at a distance of 72 km by road and 67 km by train from the District Kota. It is said that Solanki Rajputs founded it during 14th or 15th century. The place derives its present name because it is believed to have been populated by the inhabitants of twelve (Barah) adjacent villages. Baran District is divided into two tehsils (Shahabad and Kishanganj). The number of villages in Baran District is 439. The Saharia are the only Primitive Tribe of Rajasthan and inhabit in the Shahabad and Kishanganj tehsils of Baran District. Shahabad tehsil is at a distance of 80 km from Baran District and Kishanganj tehsil is at a distance of 40 km from Baran District. The total area of Baran District is 2898 sq. km. Literacy Rate in Baran District is 38.4. Shahabad tehsil has a total area of 1469 sq. km whereas Kishanganj tehsil has an area of 1429 sq. km. The number of villages in Shahabad tehsil are 236, out of which only 146 are inhabited villages and rest of the villages are uninhabited. The total population in Shahabad tehsil is 83,028. The population of Saharia in Shahabad tehsil is 28,632, which forms 34.72 per cent of the Saharia population. The population of Saharia in Kishanganj tehsil is 19,190, which forms 17.71 per cent of the Saharia population. The number of villages in Kishanganj tehsil are 203, out of which only 171 are inhabited villages. The District Baran contributed 1.84 per cent to the total population of the State. Density of population of Rajasthan works out to be 129 sq. km whereas it is only 116 persons per sq. km in case of Baran District. To cater to the needs of Saharia, special programmes were started in, 1977-78 for the development of Saharia residing in Kishanganj and Shahabad Panchayat Samitis of Baran District. According to Saharia Development Project Area (1991), the total population of Saharia in Baran District is 47,822.

The Saharia of Rajasthan are concentrated in two tehsils of Shahabad and Kishanganj which also coincide with their respective Panchayat Samitis. The Shahabad and Kishanganj tehsils can be divided into two distinct divisions (1) Upreti or Upland and (2) Taleti or Lowland. The Upreti constitutes Kishanganj teshil and the western part of Shahabad tehsil. The remaining part of Shahabad tehsil comes under the classification of Taleti.

Climate

The District Baran has a dry climate. The cold season lasts for about three and a half months from mid-November to end of February. Like almost all other areas of State, January is the coldest month. The period from March to about the middle of June constitutes the hot season. With the advance of the south-west monsoon, the temperature falls but increases steadily in the month of October when the monsoon is over. According to the data available for the year 1990, the District recorded a maximum temperature of 45°C and a minimum of 9°C. With the exception of south-west monsoon season, the air is generally dry. From March onwards uptil June which are the summer months, the relative humidities are very low in the afternoons.

River, Canals and Waterways

The principal river of the District is Chambal which has a perennial flow. Its tributaries, namely, Parwati, Kali Sindh, Parwan and Andheri along with other rivulets as Chaken, Kunu and Ahu provide an important source of water. This water is used by the Saharia for drinking and bathing purpose.

Educational Facilities

The data pertaining to educational institutions for the year 2000 revealed that while the number of colleges imparting general education in Baran District was five, it was two in the case of those relating to professional education. As for higher secondary and secondary schools, the number was 106. Among other types of institutions, there are 250 middle schools and 982 primary schools. There was, however, only one professional school in District in the year 2000. Colleges providing professional education were attended by 1,393 boys and 387 girls. As for higher secondary and secondary schools, the respective number of boys and girls was 34,950 and 14,041. In middle schools, the strength of male students was 52,995 as against 12,969 of females. Besides there were 76,792 boys and 29,019 girls in primary schools. The only professional school in the District has 42 boys and 55 girls on its rolls, bringing the aggregate of the District to 172,290 boys and 65,332 girls. Literacy rate in Baran District is 38.4. It has been observed that 98 per cent villages have the facility of primary school, 22 per cent have middle school, 10 per cent have high school and only 2 per cent have higher secondary school. It is felt that middle school onwards the educational facilities need to be strengthened throughout the District.

Medical and Public Health Facilities

Among the various type of allopathic medical institutions functioning in the District, government dispensaries claimed the largest number (58) followed by Primary Health Centres (11) providing facilities of indoor and out-door treatment, maternity and preventive health work during 2000. In Baran District there are two hospitals (one is a civil hospital and the other is a District hospital), four Community Health Centres (CHC), eleven Primary Health Centres (PHC) and fifty two Government dispensaries. These institutions are administered by the Medical Department, Ministry of Health and Family Welfare, Rajasthan State. During the year 2000, 46,647 indoor patients were treated in medical institutions out of which 15,871 were cured and 8,592 relieved. Likewise the out-door patients treated at these institutions numbered 1,76,963. Table 9 indicates the distribution of 912 in-patient beds in two hospitals of the District.

Table 9. Distribution of In-Patient Beds in Two Hospitals of District Baran

S.No		*No. of Beds*
1.	Tuberculosis	30
2.	Maternity	100
3.	Ear, Nose, Throat (ENT)	10
4.	Pediatric	80
5.	Eye	24
6.	Primary Health Centres	90
7.	General	570
8.	Maternity Child Welfare Centres	8
9.	Total	912

Apart from this as many as 44,174 couples were effectively protected under Family Planning Methods during 1999-2000.

Village Profile of the Villages Under Study

Baran District of Rajasthan is divided into Shahabad and Kishanganj tehsils. The total number of villages in

Baran District is 439. Shahabad tehsil has 236 villages and Kishanganj tehsil has 203 villages. The Shahabad and Kishanganj tehsils have been divided into Upreti or Upland and Taleti or Lowland. The number of inhabited villages in Shahabad tehsil is 146 and the number of uninhabited villages is 90. The number of inhabited villages in Kishanganj tehsil is 171 and the number of uninhabited villages is 32. The present study was conducted on the Saharia of Shahabad and Kishanganh tehsil. Fieldwork was carried out in eight villages of Upreti and eight villages of Taleti and sample was collected from 150 households each from Upreti and Taleti.

Household Information of the Villages Under Study

Table 10 shows the household profile of eight villages (Upreti). The total number of households are 1543 with a total population of 8737. The total population of scheduled castes are 5154 and scheduled tribes are 3583. Table 11 shows the household profile of eight villages (Taleti). The total number of households are 1842 with a total

Table 10. Household Profile of Saharia Villages in Upreti

S.No.	*Name of Villages*	*No. of Households*	*Total Population*	*SC Population*	*ST Population*
1	Shahpur	80	432	29	403
2	Gangapura	47	481	278	203
3	Khanda Sehrol	247	1337	842	495
4	Ghensua	58	283	105	178
5	Mahodara	270	1586	1136	450
6	Dhikwani	215	1293	746	547
7	Hatri	122	844	85	759
8	Simrania	504	2481	1933	548
	Total	1543	8737	5154	3583

Source: BDO Office (2000)

population of 9245. The population of scheduled castes are 3570 and scheduled tribes are 5675 according to Block Development Office (BDO, 2000).

Table 11. Household Profile of Saharia Villages in Taleti

S.No.	*Name of Villages*	*No. of Households*	*Total Population*	*SC Population*	*ST Population*
1	Shahabad	551	2701	1203	1498
2	Kushalpura	115	565	....	565
3	Devri	575	2989	1772	1217
4	Faredua	77	426	130	296
5	Beelkhera Dang	101	504	149	355
6	Chaura Khadi	180	838	....	838
7	Pathari	138	645	189	456
8	Sandhri	105	577	127	450
	Total	1842	9245	3570	5675

Source: BDO Office (2000)

Table 12 shows that the total population in Saharia villages of Upreti and Taleti. The total population of Saharia in Upreti is 3583. The number of males are 1882 and the number of females are 1701. The total population of Saharia in Taleti is 5675. The number of males are 2996 and the number of females are 2679.

Table 12. Total Population of Saharia in Studied Villages of Upreti and Taleti

Area	*Males*	*Females*	*Total*
Upreti	1882	1701	3583
Taleti	2996	2679	5675

Centre of Administration and Remoteness of Villages

Shahabad tehsil is the centre place of the District. It is the administrative centre, and having important institutions and offices of various importance. The distance of different villages (Upreti) under study from Tehsil and Block Headquarter is given in Table 13. Table 13 shows that the minimum distance from village to Tehsil Headquarter is of the village Shahapur of Upreti (8 km) and minimum distance from village to Block Headquarter is also from the village Shahpur (8 km) whereas the maximum distance is from the village Samrania to tehsil Headquarter is 52 km and from Block Headquarter is also 52 km.

Table 13. Distance of Different Villages (Upreti) Under Study From Tehsil and Block Head Quarter

S.No	*Name of Villages (Upreti) in Shahabad Tehsil*	*Distance of Village From Head Quarter (km.)*	*Distance of Village From Block Head Quarter (km.)*
1.	Shahapur	8	8
2.	Gangapura	9	9
3.	Khandasehrol	13	14
4.	Ghesua	19	20
5.	Mahodara	23	24
6.	Dhikwani	25	23
7.	Hatri	27	28
8.	Samrania	52	52

The distance of different villages (Taleti) under study from tehsil and Block Headquarter has been presented in Table 14. Table 14 shows that the minimum distance from village to tehsil Headquarter is of the village Shahabad of Taleti (2 km) and minimum distance from village to Block Headquarter is also from the village Shahabad (3 km) whereas the maximum distance is from the village Sandhri to tehsil Headquarter is 54 km and from Block Headquarter. It is also 55 km.

Transport and Communication Facilities

Tables 15 and 16 show the distance of different Villages of Upreti and Taleti from the pucca road and

Table 14. Distance of Different Villages (Taleti) Under Study From Tehsil and Block Head Quarter

S.No	*Name of Villages (Taleti) in Shahabad Tehsil*	*Distance of Village From Head Quarter (km.)*	*Distance of Village From Block Head Quarter (km.)*
1.	Shahabad	2	3
2.	Kusalpura	4	4
3.	Devri	10	9
4.	Faredua	17	16
5.	Beelkheradang	44	43
6.	Charuakhadi	47	48
7.	Pathari	52	52
8.	Sandhri	54	55

nearest bus stop. Table 15 presents the distribution of villages (Upreti) according to the distance from nearest pucca road and bus stop. It has been found that four villages (50 per cent) among Saharia (Upreti) are connected by pucca road, two villages (25 per cent) are situated at a distance of 5-9 km from pucca road, one village (12.5 per cent) is situated at a distance of 5 km from pucca road and one village (12.5 per cent) is situated at a distance of

Table 15. Distribution of Villages (Upreti) According to Distance (Km) from Nearest Pucca Road and Nearest Bus Stop

Area	*Distance (km)*	*Distance from Pucca Road*	*Nearest Bus Stop*
	Connected by Pucca Road	4 (50.0)	—
Upreti	< 5 Km	1 (12.5)	4 (50.0)
	5-9 Km	2 (25.0)	3 (37.5)
	10+ Km	1 (12.5)	1 (12.5)

(figures in parenthesis indicate percentage)

more than 10 km from pucca road. 50 per cent villages have bus stop within distance of 5 km, 37.5 per cent villages have bus stop at a distance of 5-9 km from the pucca road and 12.5 per cent have bus stop at a distance of more than 10 km.

Table 16. Distribution of Villages (Taleti) According to Distance (Km) from Nearest Pucca Road and Bus Stop.

Area	*Distance (km)*	*Distance from Pucca Road*	*Nearest Bus Stop*
	Connected by Pucca Road	3 (37.5)	—
	< 5 Km	1 (12.5)	3 (37.5)
Upreti	5-9 Km	1 (12.5)	1 (12.5)
	10+ Km	3 (37.5)	4 (50.0)

(figures in parenthesis indicate percentage)

Table 16 shows the distribution of villages (Taleti) according to the distance from nearest pucca road and bus stop. It has been found that three villages (37.5 per cent) among Saharia (Taleti) are connected by pucca road, three villages (12.5 per cent) are situated at a distance of 5-9 km from the nearest pucca road, 12.5 per cent villages are situated at a distance of 5 km from the nearest pucca road and 37.5 per cent are situated at a distance of more than 10 km from the nearest pucca road. 12.5 per cent villages have bus stop within a distance of 5-9 km from the nearest pucca road, 37.5 per cent villages have bus stop at a distance of 5 km from the pucca road and 50 per cent have bus stop at a distance of more than 10 km.

Status of Basic Amenities Infra Structure

The villages under study lack basic essential amenities like banks, post offices, cinema houses and chemist shops.

There are only two banks in the entire tehsil of Shahabad. There is no chemist shop in the interior villages. A brief sketch of different amenities available in the villages (Upreti and Taleti) under study are presented in Tables 17 and 18 respectively.

Table 17 shows that only village Samrania has a bank, other villages donot have a bank. Except Samrania no village has a post office. Three villages (Shahpur, Gangapura and Samrania) have a cinema house and only these vilages (Shahpur, Gangapura and Samrania) have chemist shops.

Table 17. Various Amenities Available in the Villages (Upreti) Under Study

S. No	*Name of Village (Upreti)*	*Different Amenities*					
		Bank	*Co-operative Society*	*Post Office*	*Fair Price Shop*	*Cinema House*	*Chemist Shop*
1.	Shahpur	00	01	00	01	01	01
2.	Gangapura	00	01	00	01	01	01
3.	Khanda Sehrol	00	00	00	00	00	00
4.	Ghensua	00	00	00	00	00	00
5.	Mahodra	00	01	00	00	00	00
6.	Dhikwani	00	01	00	00	00	00
7.	Hatri	00	00	00	00	00	00
8.	Samrania	01	00	01	01	01	01
9.	Total	01	04	01	03	03	03

Table 18 shows that only village Shahabad has a bank, a post office and cinema house Only two villages (Shahabad and Devri) have co-operative society, fair-price shops and chemist shops.

POPULATION PROFILE

Socio-Economic and Environmental Characteristics

Table 19 shows the socio-economic and environmental characteristics of the Saharia of Taleti and Upreti. All the

Table 18. Various Amenities Available in the Villages (Taleti) Under Study

S. No	*Name of Village (Taleti)*	*Different Amenities*					
		Bank	*Co-operative Society*	*Post Office*	*Fair Price Shop*	*Cinema House*	*Chemist Shop*
1.	Shahabad	01	01	01	01	01	01
2.	Kusalpura	00	00	00	00	00	00
3.	Devri	00	01	00	01	00	01
4.	Faredua	00	00	00	00	00	00
5.	Beelkheradang	00	00	00	00	00	00
6.	Chaurakhadi	00	00	00	00	00	00
7.	Pathari	00	00	00	00	00	00
8.	Sandhri	00	00	00	00	00	00
9.	Total	01	02	01	02	01	02

Saharia houses (100%) are of Kaccha type. The houses of Saharia are constructed from stones and mud. The blocks of irregular stones are placed one above the other and plastered with mud. The roofs are made of small pieces of stone slabs plastered with mud. Some houses have thatched roofs. There are no windows in these houses. The houses of Saharia are, therefore, full of darkness even during day-time. When the meals are prepared the room fills with smoke and causes suffocation. All houses have only one room. Most of the Saharia have nuclear families (98%). 26 per cent households are landless, 59.7 per cent have less than five Bighas of land and 14.3 per cent families have six to ten Bighas of land in their possession. About 71 per cent households obtain their drinking water from the hand-pumps, 15.7 per cent from rivers, 13.3 per cent from the wells. The water tanks were found empty. 84 per cent households have no electricity. The main source of income of Saharia include agriculture (43%), forest produce (16%) and labourers in construction sites (41%). None of the households have toilet facility. All the households use earthen smoke ovens (chulhas). Wood is used for fuel in all the houses.

Table 19. Socio-Economic and Environmental Characteristics of Saharia Villages/Households (n=300)

S.No.		*Taleti*	*Upreti*	*Total*	%
1.	Type of House				
	Kachha	150	150	300	100.0
	Pucca	–	–	–	0.0
	Mixed	–	–	–	0.0
2.	Type of house hold				
	Nuclear	150	144	294	98.0
	Joint	–	6	6	2.0
3.	Land (in bighas) (1 bigha = 0.15 Hectares)				
	Landless	40	38	78	26.0
	1-5	92	87	179	59.7
	6-10	18	25	43	14.3
4.	Main source of income				
	Agriculture	62	67	129	43.0
	Forest produce	28	20	48	16.0
	Labour in construction sites	60	63	123	41.0
5.	No. of Rooms (including kitchen)				
	One	150	150	300	100.0
	Two	–	–	–	0.0
6.	Houses Electrified				
	Yes	30	18	48	16.0
	No	120	132	252	84.0
7.	Toilet Facility				
	Yes	–	–	–	0.0
	No	150	150	300	100.0
8.	Type of Chulha/Fuel used				
	Hearth smoke	150	150	300	100.0
	Wood	150	150	300	100.0
9.	Source of Drinking water				
	Hand Pump	105	108	213	71.0
	River	25	22	47	15.7
	Well	20	20	40	13.3
	Tank	–	–	–	0.0

As mentioned earlier the data was collected on 150 households each from Upreti and Taleti areas. The population of Saharia (Upreti) in the present study has been found to be 767 (403 males and 364 females) whereas in Saharia (Taleti) the population composed of 1077 (569 males and 508 females). The masculanity proportion has been found to be 52.7.

Age

A large number of respondents (38.0%) were found to be in the age groups of 26-35 years, 7 per cent respondents were found in the age groups of 36-45 years, 19.7 per cent respondents were above 45 years and 25.3 per cent respondents were found in the age group of 18-25. Table 20 shows the number of respondents according to Age in the present study.

Sex and Marital Status

Table 21 presents the number of respondents according to sex and marital status in the Villages of Upreti and Taleti. Of the respondents covered in the study, 52.3 per cent are males and 47.7 per cent are females. The majority of the respondents (92.0 per cent in Upreti and 94.6 per cent in Taleti) are married.

Table 22 shows the income of the respondents among Saharia (Upreti). 56 per cent respondents have income in the range of Rs 1000/-1999/-, 18 per cent have income in the range of Rs 500/-999/-, 16 per cent have less than Rs 500/- and only 10 percent have income in the range of Rs 2000/-2999/-.

Table 23 shows the income of the respondents among Saharia (Taleti). 42 per cent household have income less than Rs 500/-, 32 per cent have income in the range of Rs 500/-999/- and 26 percent have income in the range of Rs 1000/-1999/- only.

Table 20. Number of Respondents According to Age in the present study

Age	*Age (in years)*												
	18-25	*M*	*F*	*26-35*	*M*	*F*	*36-45*	*M*	*F*	*45 & above*	*M*	*F*	*T*
Up.	49 (32.7)	27 (18.0)	22 (14.7)	42 (28.0)	22 (14.7)	20 (13.3)	21 (14.0)	11 (7.3)	10 (6.7)	38 (25.3)	20 (13.3)	18 (12.0)	150 (100.0)
Ta.	27 (18.0)	15 (10.0)	12 (8.0)	72 (48.0)	36 (24.0)	36 (24.0)	30 (20.0)	15 (10.0)	15 (10.0)	21 (14.0)	11 (7.3)	10 (6.7)	150 (100.0)
Tot.	76 (25.3)	42 (14.0)	34 (11.3)	114 (38.0)	58 (19.3)	56 (18.7)	51 (17.0)	26 (8.6)	25 (8.4)	59 (19.7)	31 (10.3)	28 (9.4)	300 (100.0)

Figures in parenthesis indicate percentage

Note: Up.=Upreti, Ta.=Taleti, M=Male, F= Female, T= Total

Table 21. Number of Respondents According to Sex and Marital Status in the Villages of Upreti and Taleti

Area	*Sex*			*Marital Status*			
	Male	*Female*	*Total*	*Married*	*Unmarried*	*Widow/ Widower*	*Total*
Upreti	80 (53.3)	70 (46.7)	150 (100.0)	138 (92.0)	7 (4.7)	5 (3.3)	150 (100.0)
Taleti	77 (51.3)	73 (48.7)	150 (100.0)	142 (94.6)	2 (1.4)	6 (4.0)	150 (100.0)
Total	157 (52.3)	143 (47.7)	300 (100.0)	280 (93.3)	9 (3.0)	11 (3.7)	300 (100.0)

(Figure in parenthesis indicate percentage)

Table 22. Distribution of Households by Income among Saharia (Upreti)

Income (In Rupees)	*Saharia (Upreti)*	
	No. of Respondents	*Percent*
<500	24	16.0
500-999	27	18.0
1000-1999	84	56.0
2000-2999	15	10.0
3000-3999	—	—
4000-4999	—	—
>5000	—	—
Total	150	100.0

It has been found that 41 per cent Saharia have income in the range of Rs 1000/-1999/-, 29 per cent have income in the range of less than Rs 500/-, 25 per cent have income in the range of Rs 500/-999/- and only 5 per cent have income in the range of Rs 2000/-2999/- only.

Literacy Level

It has been found that Literacy rate in Upreti villages is higher (21.5 per cent) as compared to villages in Taleti

Table 23. Distribution of Households by Income among Saharia (Taleti)

Income	*Saharia (Taleti)*	
(In Rupees)	*No. of Respondents*	*Percent*
<500	63	42.0
500-999	48	32.0
1000-1999	39	26.0
2000-2999	—	—
3000-3999	—	—
4000-4999	—	—
>5000	—	—
Total	150	100.0

(6.2 per cent). Table 24 shows the literacy status among Saharia in the Villages of Upreti and Taleti as reported by BDO Office.

Table 24. Literacy Status Among Saharia in the Villages of Upreti and Taleti

Area	*Literate No.*	*Illiterate No.*	*Total*
Upreti	772 (21.5%)	2811 (78.5%)	3583 (100.0%)
Taleti	352 (6.2%)	5323 (93.8%)	5675 (100.0%)
Total	1124 (12.1%)	8134 (87.9%)	9258 (100.0%)

Source: BDO Office (2000).

Of the total respondents, 82.0 per cent are illiterate. The educated are divided into 4 categories- 4.7 per cent literate, 9.3 per cent upto primary level, 2.3 per cent upto middle and 1.7 per cent upto secondary level of education. Table 25 shows the number of respondents according to educational level.

Table 25. Number of Respondents According to Educational Level

Area	Educational Level															
	Illiterate			*Literate*			*Primary*			*Middle*			*Secondary*			*Total*
	M	*F*	*T*	*M*	*F*	*T*	*M*	*F*	*T*	*M*	*F*	*T*	*M*	*F*	*T*	
Upreti	34 (22.6)	40 (26.6)	74 (49.3)	24 (16.0)	20 (13.3)	44 (29.3)	12 (8.0)	8 (5.3)	20 (13.3)	8 (5.3)	1 (0.6)	9 (6.0)	2 (1.3)	1 (0.6)	3 (2.0)	150 (100.0)
Talti	47 (31.3)	57 (38.0)	104 (69.3)	16 (10.6)	12 (8.0)	28 (18.6)	8 (5.0)	3 (2.0)	11 (7.3)	4 (2.6)	1 (0.6)	5 (3.3)	2 (1.3)	- -	2 (1.3)	150 (100.0)
Total		178 (59.3)			72 (24.0)			31 (10.3)			14 (4.6)			5 (1.7)		300 (100.0)

Figures in parenthesis indicate percentage.
Note:- M=Male, F= Female, T= Total.

Age at Marriage

Table 26 presents the distribution of age at marriage among Saharia females. It has been found that 42 per cent of Saharia females are married between the age groups of 16-18, 38 per cent are married between the age groups of 13-15, 8 per cent are married between 10-12 and 2 per cent are married in the age groups of 25-27 respectively.

Table 26. Distribution of Age at Marriage among Females

Age Groups (years)	*Saharia Women*	
	No.	*Percent*
10-12	12	8.0
13-15	57	38.0
16-18	63	42.0
19-21	9	6.0
22-24	6	4.0
25-27	3	2.0
Total	150	100.0

Table 27 shows the distribution of age at marriage among Saharia males. It has been found that 52 per cent

Table 27. Distribution of Age at Marriage among Saharia Males

Age Groups (years)	*Saharia Males*	
	No.	*Percent*
10-12	3	2.0
13-15	9	6.0
16-18	42	28.0
19-21	78	52.0
22-24	12	8.0
25-27	6	4.0
Total	150	100.0

of Saharia males are married between the age groups of 19-21, 28 per cent are married between the age groups of 16-18, 8 per cent are married between 22-24 and 6 per cent are married in the age groups of 25-27 respectively.

DEMOGRAPHIC TRENDS

Age and Sex Composition

The age-sex composition depicts the basic demographic state of a population in a region. It substantially influences the capacity or potential for future population change. Simultaneously it itself gets influenced by past population changes, i.e. by past vectors of fertility,

Table 28. Age–Sex Composition and Sex Ratio of Saharia of Upreti

Age Groups	*Male*		*Female*		*Total*	*Sex Ratio*
	No.	*%*	*No.*	*%*		
0-4 years	56	7.3	48	6.2	104	857.1
5-9 years	48	6.2	41	5.3	89	854.1
10-14 years	43	5.6	33	4.3	76	767.4
15-19 years	32	4.1	34	4.4	66	1062.5
20-24 years	38	4.9	40	5.2	78	1052.6
25-29 years	34	4.4	38	4.9	72	1117.6
30-34 years	47	6.1	32	4.1	79	680.8
35-39 years	32	4.1	26	3.3	58	812.5
40-44 years	24	3.1	18	2.3	42	750.0
45-49 years	18	2.3	14	1.8	32	777.7
50-54 years	14	1.8	18	2.3	32	1285.7
55-59 years	10	1.3	12	1.5	22	1200.0
60-64 years	4	0.5	7	0.9	11	1750.0
65-69 years	2	0.2	2	0.2	4	2000.0
70 years	1	0.1	1	0.1	2	1000.0
Total	403	52.5	364	47.4	767	903.2

mortality, migration, nupituality operating in an area/ population. Present data on age is collected in terms of completed whole or full years. This (age-data) has been then categorized into five years groups (cohorts) instead of single year numbers. Using these standard five-year age cohorts the age and sex composition of the presently studied Saharia tribal group is presented in Tables 28 and 29. It has been calculated from Tables 28 and 29 that Saharia is a growing population as 38.45 per cent of the population has been found in 0-14 years age group. Saharia of Taleti (as calculated) show higher i.e. 40.85 per cent than the Upreti (35.07 per cent) in 0-14 years age group. It appears that Saharia of Taleti are growing at faster rate than Upreti (Figures 4 & 5).

Table 29. Age–Sex Composition and Sex Ratio of Saharia of Taleti

Age Groups	*Male*		*Female*		*Total*	*Sex Ratio*
	No.	*%*	*No.*	*%*		
0-4 years	88	8.17	78	7.2	166	886.3
5-9 years	79	7.3	69	6.4	148	873.4
10-14 years	70	6.4	56	5.1	126	800.0
15-19 years	50	4.6	48	4.4	98	960.0
20-24 years	56	5.1	54	5.0	110	964.4
25-29 years	49	4.5	39	3.6	88	795.9
30-34 years	40	3.7	30	2.7	70	750.0
35-39 years	38	3.5	36	3.3	74	947.3
40-44 years	33	3.0	31	2.8	64	939.3
45-49 years	26	2.4	27	2.5	53	1125.0
50-54 years	18	1.6	20	1.8	38	1111.1
55-59 years	14	1.2	12	1.1	26	857.1
60-64 years	3	0.2	4	0.3	7	1333.3
65-69 years	3	0.2	2	0.1	5	666.6
70 years	2	0.1	2	0.1	4	1000.0
Total	569	52.8	508	47.1	1077	892.7

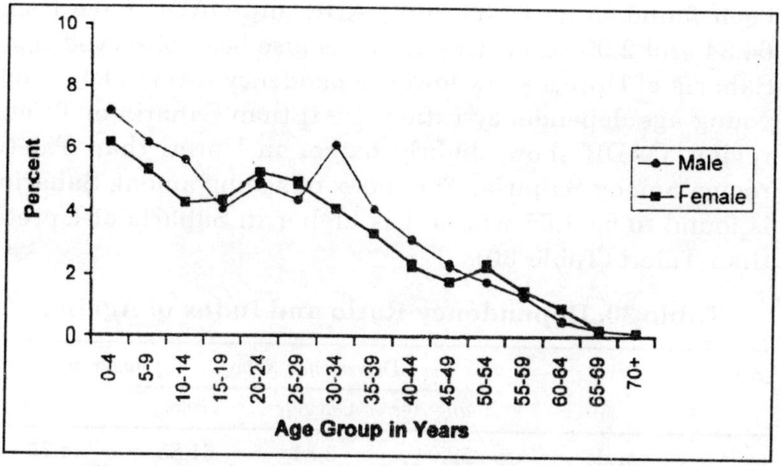

Fig. 4. Population Structure of Saharia (Upreti).

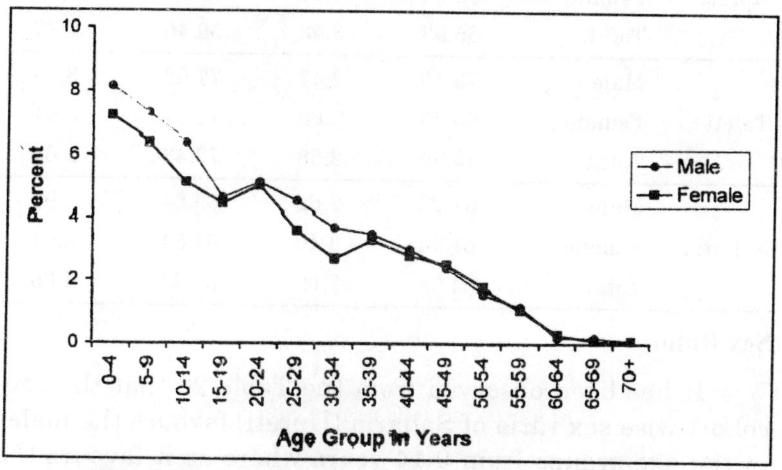

Fig. 5. Population Structure of Saharia (Taleti).

Dependency Ratio and Index of Ageing

Dependency Ratio measures the impact of age composition on the livelihood activity of a population. In the present study, all the three measures viz. young age dependency ratio (YADR), old age dependency ratio (OADR) and total dependency ratio (TDR) have been estimated. The total dependency ratio among Saharia has been found to be 67.33 and YADR and OADR have been 64.34 and 2.99 respectively. It has also been observed that Saharia of Upreti show lower dependency ratio (59.46) and young age dependency ratio (55.93) than Saharia of Taleti region. OADR show slightly higher in Upreti than Taleti region among Saharia. The index of ageing among Saharia is found to be 4.65 where it is higher in Saharia of Upreti than Taleti (Table 30).

Table 30. Dependency Ratio and Index of Ageing

Area	*Sex*	*Dependency Ratio*			*Index of Ageing*
		Young Age	*Old Age*	*Total*	
Upreti	Male	59.04	2.81	61.85	4.76
	Female	52.59	4.31	56.90	8.20
	Total	55.93	3.53	59.46	6.32
Taleti	Male	73.15	2.47	77.62	3.38
	Female	68.35	2.69	71.04	3.94
	Total	70.85	2.58	73.43	3.64
Saharia	Male	67.02	2.62	69.64	3.91
	Female	61.44	3.40	64.64	5.54
	Total	64.34	2.99	67.33	4.65

Sex Ratio

It has been observed from the Table 28 that the age-cohort wise sex ratio of Saharia (Upreti) favours the males in the age groups from 0-14 years where as it favours the females from the age groups 15-29 years. Then again it favours the males from 30 years to 49 years and favours females from 50 years onwards upto 70 years. The overall sex ratio has been found to be 903.2.

It has been observed from the Table 29 that the age-cohort wise sex ratio of Saharia (Taleti) favours the males distinctively in the age groups from 0-14 years whereas it favours the females in the age groups 45-49, 50-54, 60-64 and 70 years respectively. The overall sex ratio has been found to be 892.7 (Table 29). The sex ratio among Saharia has been found to be 897.1 (Table 31).

Table 31. Trends in Basic Demographic Indicators of Saharia of Upreti and Taleti

S. No.	Demographic Indicators	.Saharia (Upreti)	Saharia (Taleti)	Saharia (Total)
1	Sex Ratio	903.8	892.8	897.1
2	Crude Birth Rate	44. 3	27.8	34.7
3	Crude Death Rate	10.4	12.9	11.9
4	General Fertility Rate	168.3	113.2	137.0
5	Total Fertility Rate	5.04	3.52	4.21
6	Infant Mortality Rate	82.3	90.0	85.9
7	Masculinity Proportion	52.5	52.8	52.7

Crude Birth Rate (CBR)

The Crude Birth Rate (CBR) declined from 42.4 per 1000 population in 1971 to 32.1 in 1997 (NFHS–98). CBR of Saharia has been found to be 34.7/1000 which is equal to Rajasthan and is more than national average (Table 32). Saharia of Upreti show much higher CBR (44.3/1000) than the Taleti (Table 31).

Table 32. Demographic Information of Saharia in Relation to Rajasthan and India

S. No.	Demographic Indicator	Saharia (Present Study)	Rajasthan (1991)	India (2001)
1.	Sex Ratio	897.1	910	933
2.	Crude Birth Rate (CBR)	34.7	34.7	26.4
3.	Crude DeathRate (CDR)	11.9	10.4	8.7
4.	InfantMortality Rate (IMR)	85.9	83.0	70.0
5.	Total Fertility Rate (TFR)	4.21	4.2	2.6

Crude Death Rate (CDR)

It has been observed that Crude Death Rate (CDR) has declined from 16.8 per 1000 population in 1991 to 8.8 per 1000 in 1998 (NFHS- 1998). Saharia present high CDR of 11.9 than the Rajasthan (10.4) and national average (8.7) (Table 32). Saharia of Taleti show much higher CDR (12.9) than the Upreti (Table 31).

General Fertility Rate (GFR)

General Fertility Rate (GFR) has been found to be 137.0 in Saharia whereas Saharia of Upreti show higher General Fertility Rate (168.3) than Saharia of Taleti (Table 31).

Age Specific Fertility Rate (ASFR)

Table 33 presents the age-specific fertility rates of women of various age-groups. It has been observed that in both the regions of Saharia i.e. Upreti and Taleti, the age specific fertility rates have been found to be increased from the age-groups of 15-19 years to 25-29 years. There after age specific fertility rates decreased gradually showing a little and uneven trends at later years in both the regions of Saharia. The figure shows that the number of children to be expected for 1000 married women in their reproductive life has been 841.42.

Total Fertility Rate (TFR)

Total Fertility Rate (TFR) is estimated at 4.21 births per woman among Saharia which is higher than national average (Table 32). It has been observed that the Total fertility rate (TFR) is higher in Saharia of Upreti (5.04) than the Saharia of Taleti (Table 31). It indicates that ever married Saharia women may expect 4.2 children in her life-time.

Infant Mortality Rate (IMR)

It is note worthy that Rajasthan's Infant mortality Rate (IMR) has declined from 123 per 1000 live births in

Table 33. Age–Specific Fertility Rate among the Saharia of Upreti & Taleti

Age Groups	*Total No. of Saharia Women*		*No. of live Birth During last year*		*Age Specific Fertilitty Rate*		
	Upreti	*Taleti*	*Upreti*	*Taleti*	*Upreti*	*Taleti*	*Total*
15-19	34 (16.8)	48 (18.1)	4	6	117.64	125.0	121.95
20-24	40 (19.8)	54 (20.3)	9	10	225.0	185.18	202.13
25-29	38 (18.8)	39 (14.7)	14	9	368.42	230.77	298.70
30-34	32 (15.8)	30 (11.3)	3	2	93.75	66.67	80.65
35-39	26 (12.8)	36 (13.5)	2	1	76.92	27.78	48.39
40-44	18 (8.9)	31 (11.6)	1	1	55.56	32.25	40.82
45-49	14 (6.9)	27 (10.1)	1	1	71.42	37.03	48.78
Total	202	265	34	30	5.04	3.52	4.21

figures in parenthesis indicates percentage

1971 to 83 in 1998- a 33 per cent decline (NFHS-98). Saharia indicate an IMR of 85.9 which is higher than Rajasthan (83.0) and Indian national average (70) (Table 32). Saharia of Taleti show much higher IMR (90) than Saharia of Upreti (Table 31).

Health Infrastructure

In Baran District there are two hospitals. One is a civil hospital (allopathic) and the other is the District hospital. The Medical Department of the State administers these hospitals. The organization set up of health infrastructure by allopathic system of medicine is given in Figure 6. The organization operates from health secretariat at Baran. At the district level the Chief Medical Health Officer (CMHO) look after District Health and

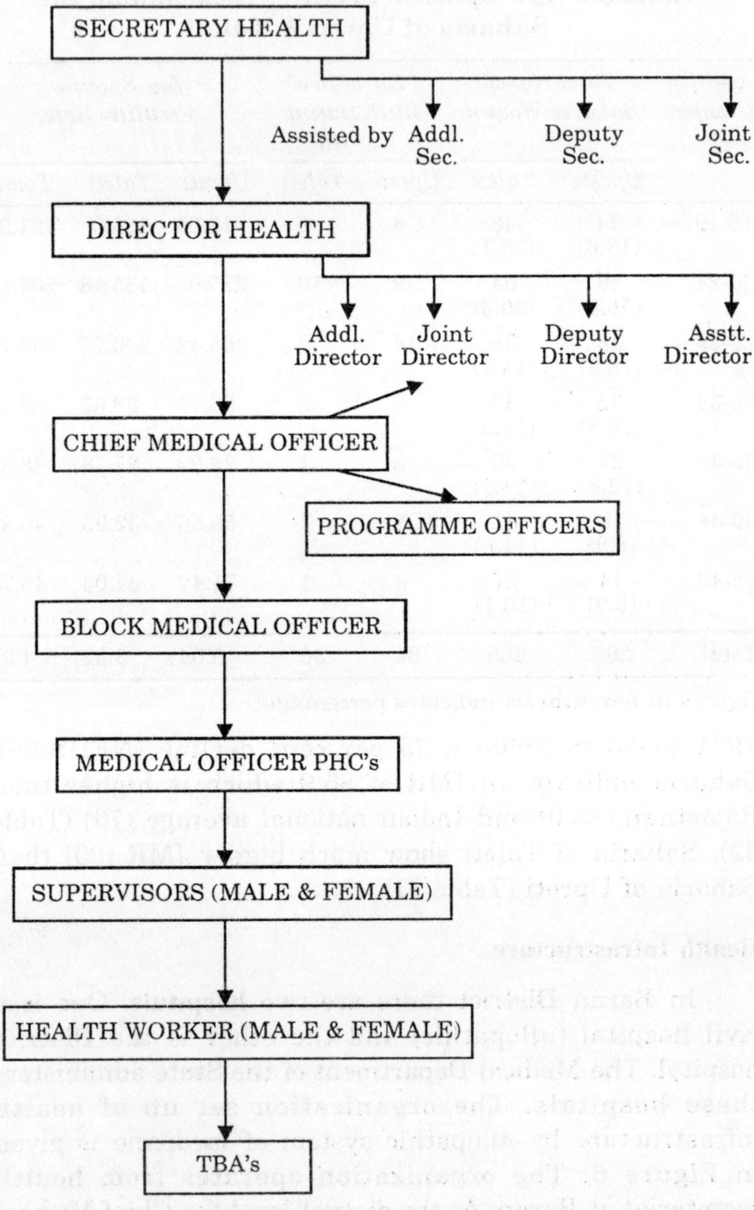

Fig. 6. Health care System Facilities-Rajasthan (Allopathic).

Family Welfare Programmes alongwith the various National Health Programmes, like TB, leprosy, Goitre etc. Besides controlling the District hospital, the Chief Medical Health Officer (CMHO) is also the main Controlling Officer of all the medical and para-medical officers employed in Government Institutions such as Community Health Centres (CHCs), Primary Health Centres (PHCs) and Sub-centres (SCs).

At Block level the Block Medical Officer (BMO) is the main functionary who is responsible for executing Health and Family Welfare Programmes as well as other marginal health programme.

At grassroots level, the health supervisor and the health worker are attached to the PHC and Sub-centre respectively and are the main workers who provide preventive and curative medical help to villagers.

There is one civil hospital and one District hospital, four CHC's, eleven PHC's, thirty-two Sub-Centers and fifty-two Government dispensaries in the District. Althought significant studies have been made in the field of health and education there is much to be achieved in this area of welfare activities.

If one glances through the records of vacant position of doctors and paramedical staff in Baran, one finds that against 10 sanctioned specialist positions, eight are vacant. Positions of five doctors are vacant in the eleven PHC's. Similarly, eighteen of the forty-eight Female Health Worker's (FHW) positions are lying vacant. These positions are extremely important for delivering the health care services at the grass root level. Table 34 shows that many vital positions are lying vacant. The status of the rural health infrastructure is not satisfactory. There are hardly any laboratory and diagnostic facilities in the PHC and Sub-Centres.

Table 34. Existing Health Infrastructure in District Baran

Sl. No.	*Category*	*Sanctioned*	*In Position*	*Vacant*
1.	Chief Medical Officer (CMO)	01	01	
2.	Block Medical Officer (BMO)	03	02	01
3.	Doctor	44	39	05
4.	Specialist	10	02	08
5.	Surgeon	06	03	03
6.	Ward Sister	06	02	04
7.	Nurse	36	11	25
8.	Pharmacist	30	07	23
9.	ANM	06	02	04
10.	Female Health Supervisor	10	07	03
11.	Male Health Supervisor	09	04	05
12.	R/Grapher	05	03	02
13.	O. Pth. Assit.	04	02	02
15.	Female Health Worker (FHW)	48	30	18
16.	Health Worker Male	34	22	12

Morbidity Pattern/Burden of Diseases

Information gathered from Allopathic District Hospital Baran regarding morbidity gives following picture for the year 1999-2000 and the data was also calculated from 199 patients of Saharia (104 and 95) (Table 35) and detailed analysis has been given in the subsequent pages.

It can therefore be seen that majority of cases belong to tuberculosis (60.8%), skin diseases (scabies & eczema) (14.0%), malaria (3.5%), typhoid (4.0%), eye diseases (4.0%), mental illness (3.0%), diarrhea (3.0%), asthma (2.0%), Sexually Transmitted Diseases (STDS) (2.5%) and others (3.0%) are diseases that occur regularly among Saharia.

The Medical doctors reported that due to poverty some Saharia women are tempted to indulge in physical and

Table 35. Morbidity Profile, District Baran (April 1999 through March 2000) and Saharia

S.No.	*Ailments*	*No. of individuals who had taken treatment (Baran District)*	*Saharia (Present Study)*
1.	Tuberculosis (TB)	10347 (22.18%)	121 (60.8%)
2.	Skin Diseases	7427 (15.92%)	28 (14.0%)
3.	Malaria	227 (0.48%)	7 (3.5%)
4.	Typhoid	1387 (2.9%)	8 (4.0%)
5.	Eye Diseases	6337 (13.58%)	8 (4.0%)
6.	Mental Illness	109 (0.23%)	6 (3.0%)
7.	Diarrhea	5081 (10.82%)	6 (3.0%)
8.	Asthma	4487 (9.6%)	4 (2.0%)
9.	STDS	889 (4.9%)	5 (2.5%)
10.	Others	10356 (22.2%)	6 (3.0%)
	Total	46647 (100%)	199 (100.0%)

Source: District Hospital, Baran (Annual Report)

unsafe sex with the strangers and carry the infection back home. This is the cause of Sexually Transmitted Diseases (STDs) among Saharia. However no HIV positive case is detected in the District. Most vulnerable groups are adult males and children. The incidence of tuberculosis is very high among the Saharia (60.8%). The reasons for this are poor nutrition, unhygienic living conditions, poor environment surroundings, smoking and drinking habits. One can frequently see the children playing in mud, they are very susceptible to bacterial and fungal infections.

Table 36 presents the distance of the villages (Upreti) from nearest health facility. Majority of the villages (50 per cent) have Sub-Centre within the range of 5 km, 25 per cent villages have Sub-Centre within the villages and 25 per cent villages have Sub-Centre within the range of 5-9 km. 12.5 per cent villages have Primary Health Centre

(PHC) within the range of 5 km, 12.5 per cent have PHC within the village and 50 per cent villages have PHC at a distance of 5-9 km from the village and 50 per cent villages have PHC at a distance of more than 10 km. One village (12.5 per cent) is having Block PHC, 25 per cent villages have Block PHC within 5 km, 12.5 per cent villages have Block PHC within 5-9 km, 50 per cent villages have Block PHC situated at a distance of more than 10 km from the village. All the villages (100%) have a District hospital situated at distance of more than 10 km.

Table 36. Distribution of Villages (Upreti) According to Distance (Km) from Health Facility

Area	*Distance (km)*	*Sub- Centre*	*Primary Health Centre*	*Block Primary Health Centre*	*District Hospital*
Upreti	Within Village	2 (25.0)	1 (12.5)	1 (12.5)	—
	< 5 Km	4 (50.0)	2 (25.0)	2 (25.0)	—
	5-9 Km	2 (25.0)	1 (12.5)	1 (12.5)	—
	10+ Km	—	4 (50.0)	4 (50.0)	8 (100%)

(figures in parenthesis indicate percentage)

Table 37 shows the distance of the villages (Taleti) from nearest health facility. Majority of the villages (50 per cent) have Sub-Centre located at distance of more than 10 km from the village, 25 per cent villages have Sub-Centre within the village, 12.5 per cent villages have Sub-Centre within the range of 5 km and 12.5 per cent villages have Sub-Centre within 5-9 km. 75 per cent villages have Primary Health Centre (PHC) at a distance of more than 10 km from the village, 12.5 per cent have PHC within the village, 12.5 per cent villages have PHC within 5 km. One village (12.5 per cent) is having Block PHC, 75 per cent villages have Block PHC at a distance of more than

10 km, 12.5 per cent villages have Block PHC within 5 km. All the villages (100%) have a District hospital situated at a distance of more than 10 km.

Table 37. Distribution of Villages (Taleti) According to Distance (Km) from Health Facility

Area	*Distance (km)*	*Sub-Centre*	*Primary Health Centre*	*Block Primary Health Centre*	*District Hospital*
Taleti	Within Village	2 (25.0)	1 (12.5)	1 (12.5)	—
	< 5 Km	1 (12.5)	1 (12.5)	1 (12.5)	—
	5-9 Km	1 (12.5)	—	—	—
	10+ Km	4 (50.0)	6 (75.0)	6 (75.0)	8 (100.0)

(figures in parenthesis indicate percentage)

HEALTH SEEKING BEHAVIOUR

Concept of Health

Saharia believe in supernatural powers and they have strong feeling that different deities and spirits are responible for causation of different diseases. The spirits of dead ancestors are also believed to play an important role in ensuring good health, prosperity and protection to the community. Among Saharia, there is a strong belief that persons emanating evil, having mystical powers like evil eye and evil touch can also affect or deteriorate the health of a person. There are traditional healer ('Janteyar') in the community who identify the cause of illness and treatment is based on it. The traditional healer ('Janteyar') develops faith and assurance for good health in the patient and treats him. Saharia do not associate disease and epidemics with their insanitary condition of living but associate them to the wrath of the Gods and Goddesses. Saharia have the knowledge that when a man starts losing his appetite ('Bukha'), his health begins to deteriorate. If a man has good appetite, his health remains sound. Loosing appetite is considered bad for the health. Physical isolation, ecological legacy, traditional culture and beliefs have made the Saharia self-reliant and they have developed protective mechanism for their survival. Majority of Saharia especially in the interior villages are ignorant about the modern amenities and still follow the rituals of pleasing supernatural powers.

Concept of Illness

When a person could not move, is bed ridden, has physical discomfort, has loss of appetite, is unable to peform daily activities, is having certain physiological disorder, a person is considered to be ill by majority of the Saharia. The Saharia are basically rural people and they have to do a very hard manual labour in difficult terrain, forest and hilly areas for their survival. A man or woman is considered sick if he or she is not able to do physical labour and withdraws himself from the work assigned. Saharia also believe that certain diseases take time to reveal physical mal-functioning.

SOCIAL/ENVIRONMENTAL CONDITIONS

It has been observed that the individual's health is strongly influenced by their environmental conditions and lifestyle and that these factors produce marked inequalities of health.

Food Habits

In the past, food habits of the Saharia were fully wild and primitive. Various forest products and flesh of different animals formed their food. They were carnivorous. Among the forest products, which the Saharia used in the form of food include mahua, chirongi, gum, honey and amwala. The grains which Saharia used to consume in the past were of inferior quality from the point of view of nutrition. Now the Saharia eat Jawar, Bajra and Wheat. But the consumption of cereals, pulses and vegetables is very less, mainly because of low production and the cost factors.

Housing

Warmth and Shelter are among the most basic needs of humans. People living in damp housing are much more likely to develop asthma and respiratory infections. This is particularly true for children. Crowded housing increases the likelihood of transmission of infection between residents. The houses of Saharia are constructed from stones and mud. The blocks of irregular stones are placed one above the other and plastered with mud. The houses have thatched roofs. The houses are generally rectangular. There are no windows in the houses. The houses of Saharia are filled with darkness even during day time. When meals are prepared the room fills with smoke and causes suffocation. It has been observed that the conditions under which a Saharia lives have a major effect on how healthy they will be.

Houses of Saharia do not have any arrangements for light and air in the villages. Poor lighting during night hours, absence of suitable exit for smoke, i.e. too much smoke in crowded and ill-ventilated house may be the cause for the high incidence of 'ratondi' (night blindness). Non-availability of fresh air during night causes respiratory troubles. The traditional customs, ignorance (illiteracy) and poverty, on the one hand and indifference and reluctance of the management in these regards on the other hand are the main causes for their living in this condition.

Sanitation and Hygiene

Sanitation among the Saharia is very poor. It has been observed that the Saharia go to the field in the morning and evening for defecation. The existing habits of defecation may be responsible for ill health of Saharia in those areas. The medical doctors at PHC mentioned that the germs of cholera, typhoid and dysentry are adequately present in the stool of affected men. During rainy season these germs get mixed with stream water and this contaminated water is used by the Saharia and thus they suffer from these diseases. These parasitic infections can be controlled by providing sanitary facilities to the Saharia.

Personal hygiene is also quite poor among the Saharia. This is due to poor awareness about the importance of personal hygiene. The state of environmental and personal hygiene was found very deplorable during the fieldwork. Due to poverty and illiteracy the Saharia are unaware and ignorant about the fatality of such poor environmental sanitation.

Drinking Water Supply

Many infectious diseases such as cholera, typhoid and polio are carried in water. The organisms that cause these diseases can get into drinking water if human excreta are not properly disposed of. It is needless to emphasize the necessity of developing safe and hygienic drinking water in each and every village of the tribal area, where high infestation in almost every village results into diverse health problems, untold miser and suffering and loss of considerable manpower. The principle source of these infections is the pollute drinking water from village ponds, rivers, hand pumps, tanks etc. On account of the rocky terrain and lack of facilities for hygienic drinking water, people have to fall back upon the rain water collected in ponds, rivers and the hand pumps. The water is taken directly from rivers, wells and hand pumps for consumption. During summer the hand pumps and wells get dried and water is collected in water tanks. Water crisis in the interior villages becomes acute during summer.

There was recent outbreak of infectious diarrhoea among children of Sandhri Village. This was caused due to contaminated water. According to an article in the newspaper 'Patrika' dated 26 Sept. 1999, two children died in the village Sandhri due to diarrhoea and condition of 52 children was found critical. As told by the 'Patel' (headman) the case of diarrhoea, was reported on 25th Sept.1999 by a school teacher. Two Saharia girls aged 7 years and aged 3 years died in village Sandhri due to diarrhoea. Then the Sub Divisional Officer (S.D.O.) was informed immediately about the situation in Sandhri. A team of doctors from Primary Health Centre (PHC) Shahabad then rushed to the village Sandhri.

Smoking and Drinking Habits

Table 38 presents the number of respondents who smoke and drink. It has been observed that the majority of respondents smoke (93.3%) and drink (86.0%). They use tobacco and locally available 'Tendu' leaves (Diospyros melanoxylon) for smoking in the form of 'Bidi' while locally prepared alcoholic drink using some forest fruits and 'Mahua' flowers (Madhuca indica) is much popular, there is no restriction on females either for smoking or drinking along with their spouses.

PATTERNS OF TREATMENT

Ethnomedical System (Traditional Medical System)

The elements of primitive medicine are still practiced in many societies, both in the less developed and advanced regions, in one form or the other. The tribals in India believe that most of the diseases, specially epidemics, are caused by evil spirits. They also believe that any breach of taboo will anger the clan god and the culprit will be punished. In fact, the tribals associate the disease to sorcery, breach of taboo, object intrusion, spirit intrusion and soul loss (Majumdar, 1958). To protect from the wrath

Table 38. Incidence of Smoking and Drinking (Local) Among Saharia Respondents.

Area	Alcoholic Drink		Smoking		Total
	Yes	No	Yes	No	
Upreti	128 (85.3)	22 (14.7)	138 (92.0)	12 (8.0)	150 (100.0)
Taleti	130 (86.6)	20 (13.3)	142 (94.7)	8 (5.3)	150 (100.00
Total	258 (86.0%)	42 (14.0%)	280 (93.3%)	20 (6.7%)	300 (100.0%)

Figures in parenthesis indicate percentage

of supernaturals, the tribals worship benevolent and protective deities and spirits. In case of diseases supposed to have been caused by evil eye, sorcery or witchcraft, they seek the help of the traditional magician-cum-witch-doctor in the community.

In the tribal society, the social and cultural activities revolve around gods and spirits. They have specific gods for each and every activity of life. Likewise, they have gods for health, disease and calamities. They worship these deities to protect them from evil spirits, diseases and epidemics. For example the Saharia worship Tejaji, a deity related with snakes. Tejaji protect them from snakes and save their life in forest.

Like other tribes in India the Saharia, have also erected structures, though not as elaborate, conspicuous and posing as found among the non-tribal rural communities to house the images of various kinds of supernatural beings. Both at the individual as well as at the community level, the Saharia realise the responsibility of a safe placement of such images. Represented through various symbols and figures of various forms and material, they are kept at one or different places in and around the tribal settlements or in the near proximity of natural surroundings. Such religious places remain open to one and all irrespective of age, sex, etc. Man's interaction with such supernaturals does not merely end in the preservation of symbols and images. Rather, it is seen that they are worshipped and appeased on specific occasions in the year. Offerings of various kinds, sacrifice etc., are made to them on specific occasions. The will of community holds strong to keep up such values and traditions.

The Saharia believe that these supernaturals if disturbed through any gesture or forgetfulness, can cause severe condition that would prove damaging to the community health. The Saharia try to appease these supernaturals by offering prayers, worshipping them regularly and also on specific occasions like religious festivals, marriage ceremonies and rituals etc. The Saharia have their own explanations of sickness on the basis of which they develop their peculiar ways of curing diseases. This ethnomedical system is in practice to a considerable extent especially among the Saharia who live in interior areas.

The Saharia believe in the presence of malevolent and benevolent powers. These supernatural powers cause sickness to individual or community on ignorance or negligence of any sort by the tribal. To calm down the wrath and anger of these supernatural powers, the tribals have discovered different magico-religious methods. By propitiation and prayer, they believe, a cordial relationship between an individual and supernatural power can be established and human sufferings can be reduced or cured. In addition to magico-religious method of treatment, the Saharia also use various natural remedies, decoction of herbs and healing lotions.

The Saharia believe that diseases and misfortunes are caused by hostile spirits, ghosts of the dead or by the use of evil spiritual powers by enemies. That a disease which is caused by evil spirits can only be cured by spiritual powers, is the main idea behind their system of treatment. Therefore, for all ailments they go to their traditional healer locally called 'Janteyar'. All diseases like typhoid, tuberculosis, tetanus, pneumonia, fits, measles are considered to be caused by evil spirits and ghosts of the dead. They believe in the hierarchy of deities. Typhoid is believed to be caused by a senior deity so it is called as 'Bade Maharaj'. Tuberculosis or tetanus is caused by the angry soul of forefathers.

Tetanus ('Dhanushban') is believed to be caused by evil spirits and ghosts. While working in forests, the Saharia get hurt by the axe which often causes tetanus. Pneumonia, called as 'Dibbirog' by Saharia, is considered

Table 39. Diseases caused by Wrath of Deities/Ghosts.

S.No.	Class of Disease Causation	Disease Associated	
		Local Name	*English Name*
1	Wrath of gods (Devta) or goddesses (Devi)		
	(a) Badi Mata	Badi Mata	Measles
	(b) Sitla Mata	Sitla Mata	Chicken Pox
	(c) Amba Mata	....	Unsafe Delivery
	(d) Tejaji	....	Snake Bite
	(e) Hanuman, Jind, Thakur	....	Fever
2	Evil spirits and ghosts of the dead	Bade Maharaj Chotte Maharaj Dibbirog Dhanushban	Typhoid Malaria Pneumonia Tetanus
3	Evil Eye or Nazar Lagna or Dakini Lagna	Dakini Lagna or Bhoot Lagna or Upri Hawa	Fits

to be caused by the attack of 'Kaccha Mashan', ghost of those children who died while taking milk. For treatment of such diseases, the Saharia go to a 'Janteyar' (traditional healer). A 'Janteyar' represents a deity. He also checks the 'Nabj' (Pulse) of the patient. The 'Jantyar tells the patients to make offerings of coconut, clarified butter (ghee), sugar wafers (patasha) and jaggery on the platform of 'Bajrang' or 'Thakur' or 'Jind'. Incense sticks are also lighted on the stone platforms to please these gods. Consumption of liquor, meat and oil in the vegetables is avoided during the period of sickness. When the deity appears through him, his body starts shivering rapidly, he goes in trance and then he answers all the queries and tells the cause of ailment and suggest remedies. He also gives 'spiritual ash' to the patients. Restrictions on having certain foods are also imposed by 'Janteyar'. The Saharia have specific gods for their health and disease, for calamities and so on. Diseases caused by wrath of deities are listed in Table 39.

Sacred Beliefs and Practices

The idols of deities are kept on an elaborate construction or a raised square platform often below an auspicious trees like Ficus benghalensis, Ficus religiosa etc. The fauna and flora of the Shahabad region enjoy functional and spiritual importance among the Saharia tribe as their plant/animal deities, to terms, providing habitat for their shrines, material for their offerings and rites and rituals. Animals deities are also worshipped by Saharia with a conviction that their appeasement would provide protection from the representatives associated, e.g. Tejaji (snake deity), Bhairav, Baba deo etc. Plants too are held sacred and worshipped as deities, some important examples mentioned below will elucidate.

1. Ficus benghalensis (Bargad)

The tree has religious signifigance and is worshipped

both by men and women for fulfillment of wishes. Saharia construct a platform around the external roots of this tree.

2. Ficus religiosa (Pipal)

There is a myth associated with the origin of this tree. It is said that once Goddess Parvati became very angry with Vishnu and Brahma and cursed them. As a result, the first Pipal tree was created from the body of Vishnu and the first palash tree from the body of Brahma. The Saharia believe that God Vishnu reside in this tree. The Saharia worships the tree and destroying it is considered a sin. During the day of Dasha Mata celebration (in the month March) womenfolk move around the tree five times and tie a thread around it's trunk. The tree is worshipped for bringing good fortune and for eternal marital happiness. Persons with a form of pollution e.g., in drunken state, women in menses are forbidden to go in its proximity.

Totem and Taboo

Totem and Taboos are an integral part of Saharia life. They are strictly followed and in case of failure, negligence or mistake they are punished by the society. Rigid restrictions and avoidances came to be associated with the totem i.e. the totemic taboos. During night, passing of women under the tree of Imli (Tamarindus indica) and Peepal (Ficus religiosa) is strictly prohibited. The Saharia belonging to Parondiya and Kaluj clans regard Banyan tree as their totem. Saharia (belonging to Parondiya and Kaluj clans) do not cut Banyan tree and cutting of Banyan tree is considered a taboo. Saharia belonging to Raichohan, Solvia and Garwar clans consider Asapal tree as their totem. The totems are regarded by the Saharia with awe and fear. The occurance of diseases are contributed to their wrath. The Saharia are aware of the mechanism of warding off the dangers resulting from the violation of a taboo. This involves the acts of penance and ceremonies of purification. For this purpose they recite the name of the totem, burn the incense and offer sweets to please the deity and later the sweets are distributed to the children. The totems are held in high esteem by every Saharia family and usually images of the totem are depicted either on paper or on walls and are worshipped every morning and evening by the elders of the family. The other occasions on which totems are worshipped are the death and marriage ceremony. During the period, of pregnancy the Saharia woman is required to observe a number of taboos. For example, she must not look at the Sun or the Moon during eclipses failing to do it would disfigure the baby in the womb. The Saharia believe that if an eclipse is watched by a pregnant woman then the child born will be blind/disabled. The pregnant ladies are prohibited to go to the vicinity of Akra plant for fear of being affected by evil spirit. A piece or root of Oont-kateli is placed on the navel of pregnant lady for a few minutes for easy delivery before parturition time or during pains. The pregnant ladies are not allowed to visit the Pilu or Khara-jal tree at noon (12'o clock) and in the evening, believing that evil spirits live on the tree. During menstruation women are not allowed to prepare meals and take part in religious and social ceremonies. If the delivery is delayed, offerings are made to family goddess known as Kodia Devi. The old women of the family or neighbourhood help in the delivery. A Bhangan (sweepress) is called to cut the umbilical cord (Nal Katna). The Saharia consider that it is only the work of the Bhangan (sweepress). In the interior villages there is no sweepress and the sweepress is called from other village which requires two to three days and this has and adverse effect on the health of the mother and the newly born baby. The Bhangan cuts the umbilical cord with blade and knife. This is considered to be healthy for the development of the child. The blessings of the totem are sought for the departed soul to find a place in heaven. In the marriage ceremony, image of the totem made of the flour, or drawn on the paper or walls is worshipped at every stage. After the marriage, the bride and the bridegroom pay their tributes to the totem by offering rice and vermilion. During the Holi, Deepawali and Navratra festivals the Saharia worship their totems along with other Hindu Gods.

Evil Eye or 'Najar Lagna'

The evil eye phenomenon is one of the most dreaded faith in the Saharia society. This is the act practiced by an enemy, an ill-wisher, a witch or an inadvertent act of a jealous person. The attractive health of male, beauty of a female, smiles of a charming child, happy and prosperous state of a family, much lushing crop or a tree, much more milk giving cattle may be drastically "affected by an evil eye" (Shrivastava and Joshi, 2001). The 'Janteyar' ties a black thread around the neck and forearm of the children to protect them for evil eye. The amulet ('Tabij') is made from the seeds of Aak-phorni-ki bel. The 'Janteyar' also prepared necklaces from the fruits of Putranjwa or Putjia plant, which are put around the neck of the children to ward of the evil eye. The 'Janteyar' uses twigs of Akra plant to perform 'Jhada' to the victims of evil eye. It has been observed that in cases of mental illness the Saharia always approach the 'Janteyar'.

For some ailments, the Saharia resort to herbal medicine, which they prepare themselves with the help of various plants, found in the forest. It has been observed that health and treatment are very much connected with the forest ecology. The traditional health care system and treatment are based on their deep observations and understanding of nature and environment. The various plants, which are used for medicinal value, are listed in Table 40.

Table 40. Plants Used for Medicinal Purpose.

S.No.	Botancial Name	Local Name	Medicinal Use
1	Azadirachta indica	Neem	Leaves mixed with basil leaves (tulsi) in hot water are given to patients suffering from stomachache and eye infection. The ground paste is applied on the eyes.
2	Zizyphus jujube	Bor	Branch used for brushing teeth, for toothache with gargling in hot water.
3	Saccharum spp.	Pila Ganna	Its pith is ground and drunk in warm water. It is used to cure (Bukhar) fever.
4	Cactus spp.	Nagphani	Fruit is ground in mustard oil and is applied on wounds.
5	Datura Stromium	Datura	Leaves are prepared in form of crude cigarettes (biri) enclosing the seeds and its smoke inhaled by patients of asthma. Also consumed by patients suffering from itching (khujli) and cholera (haija).
6	Basis latifolia	Mahua	Paste of its bark is applied for curing ulcers and its oil is massaged for curing skin diseases (charam rog). Ground with lemon and ginjer (adrak) in water is drunk for lu (heat stroke)
7	Acacia nilotica	Babool	Leaves dipped in warm water are put on the eyes of patient suffering from conjunctivitis.
8	Curcuma domestica	Haldi	Turmeric (haldi) solution is given to patients suffering from skin diseases (scabies and eczema). Its paste mixed with lime is applied to patients suffering from bone dislodgement or fracture.

(Table Contd...)

Table 40. Contd...

S.No.	Botancial Name	Local Name	Medicinal Use
9	Cactus spp.	Thur	Used to cure patients suffering from cough and cold attributed to air (thanda pawan). In order to decrease the cold element, heat treatment is given by burning the wood of cactus, inhaling its smoke mixing its ash (vibhuti) with salt and drinking it in a solution of warm water. The milk (latex) of thur is put in the ears to heal earche.
10	Sterculia urens	Kandol	Leaves are crushed into paste to be used as an ointment for chapped/cracked skin.

Indigenous Medicine

Indigenous medicines used for treating various ailments are given below:

(a) Name of Disease: Bukhar (Fever)

Symptoms: Body temperature, perspiration, bodyache, weakness and the patient does not like to eat or drink anything.

Therapy: Decoction of fresh leaves of Hansraj is taken with black pepper to cure fever. Juice of leaves of Bajradanta is taken in fever. Root extract of Akra or Aak is given to drink to cure fever. Seed-powder of Kumudini is also given in fever. The patient is served honey and black pepper. The patient is also served chiraita. In the case of Malaria, which is very common during and after the rainy season, the treatment sought is the same. Paste of flowers of Neem is taken orally to cure malarial fever.

(b) Name of Disease: Saradi-Khansi (Cough and Cold)

Symptoms: The tendency of Nausea persists, cold, sneezing.

Therapy: The tea prepared adding Tulsi leaves, black pepper and ginger is widely taken to cure cough, cold and fever. Stem bark of Sarifa is chewed to get relief from cough and cold. Powder of Akra flowers is taken with black pepper to cure it. The infusion of Oont-kateli plant is also taken to cure cough and cold. The fruit of Hare plant is roasted in fire and served to the patient. The fruits of Amala and Bahera are prescribed to eat. The juice of leaves of Bare, Bahera is prescribed to take. Mahua beer is also taken. Pills prepared with catechu, black pepper and clove are taken orally. Fresh leaves of Hansraj are also taken. The extract of Bair, Amarud, Jamun, Bel fruits is prepared and served to the patient. Decoction of leaves of Lalbanbasi is taken. Fresh leaves and roots of Vasak are taken with ginger.

(c) Name of Disease: Sir Darda (Headache)

Symptoms: Irresistible pain in the head, whole head or one side of the head may ache.

Therapy: During winter season, the headache is believed to be caused by cold. During summer, it is believed to be caused by heat. The oil of Til leaves, oil of Koinaha (Mahua seed), root-powder of Kanthor, is also applied on the forehead. During winter, the head and forehead are kept near the burning fire. The bark of Chandan, Ashoka, Mahua, Amala are crushed together with water. The mixture is cooked in Mahua oil and than applied over the forehead. This removes the headache. The seeds of Mahua, Mango, Imali, Jamun and Amla, are crushed together in Okhali and then applied over the forehead to remove the headache. Neem oil is given to patients suffering from headache.

(d) Name of Disease: Deha Darda (Bodyache)

Symptoms: Pain in the whole body, weakness.

Therapy: oil of Mahua seeds, Til leaves is applied on the body. The bread of Mahua flower is prepared to serve the patient suffering from bodyache. The Mahua beer is applied over the body to get relief from pain. It is also served to the patient suffering from bodyache. Roots of Narkan are taken.

(e) Name of Disease: Kan Darda (Earache)

Symptoms: Pain in the ear.

Therapy: Juice of Kanda leaves, milk (latex) of Thur, 2-3 drops of leaf decoction of Kesundo, few drops of decoction of roots of Tumbo are put in the ears to heal earache. Peacock's feathers are ground, mixed with water and the solution is also put in the ear.

(f) Name of Disease: Katana (Cut)

Symptoms: While collecting foods and other minor forest produces, hunting, ploughing fields, moving in forest and on hill, cutting woods, crops and vegetables, the incidence of cut takes place.

Therapy: Paste prepared from leaves of Neem, Babul, Bair, Karanj, Gugal is applied over the cuts for healing. Paste of leaves of Uchauty is used to cure cuts. Decoction of roots of Kataili, juice of leaves of Bajar-Bel is also applied on the cuts for early healing. Latex of Dudha, Kal-Bhangra is applied on the cuts to check bleeding.

(g) Name of Disease: Jalana (Burn)

Symptoms: The incidence of burning takes place at the time of cooking or preparing Mahua, sitting near fire during winter season. Sometimes small children touch fire due to ignorance.

Therapy: Gum of Kumbat is dissolved in water and applied on burns. A paste of leaves, flowers and seeds of Neem, Bair, Babul, Karanj, Kahua and Ekawan etc, is prepared and then applied over the burnt part of the body.

(h) Name of Disease: Ghaw (Wounds)

Symptoms: While cutting wood or working in the field this Saharia gets wounded by the axe or plough.

Therapy: The barks of Neem, Babul and Karanj are boiled together and then crushed to get mixture. The wound is washed with the water left after boiling the barks. The mixture in paste form is applied over the affected area. The paste of roots, herbs, barks, leaves and seeds are prepared to cure the wounds. The roots of Regani, Baiganaha, Kataila, are dug out, washed properly, crushed with water, cooked in Mahua or Karanj oil and then applied over the wounds. The powder of rhizomes (Haldi) with hot-milk is given to injured persons for early healing of the wounds. Paste of leaves of Dhabi, juice of leaves of Bajar-Bel, Jinjanio is applied on the wounds for early cure. Crushed leaves of Bagra are applied to heal up the wounds.

(i) Name of Disease: Chinia (Diabetes)

Symptoms: Frequent urination, fast appetite, quick thirst due to dehydration, perspiration, headache, pain in the limbs and lack of blood in the body. Therapy: The barks of Neem and Karanj are boiled together and taken regularly. Gum of Kumat as a part of food is taken. Root extract of Akra, root powder of Kholi are taken with milk. Seeds of Gwar or Guar are soaked in water in the evening and the next morning that water is taken to cure diabetes.

(j) Name of Disease: Dama or Saans Ki Takleef (Asthma)

Symptoms: Difficulty in breathing, heart beats faster.

Therapy: Fresh leaves and roots of Sanjeevani are taken with ginger to cure asthma. Leaves of Datura are prepared in the form of crude cigarettes (biri) enclosing seeds and its smoke inhaled by the patients. Decoction of stem-bark of Khair is taken at bed- time to cure asthma. Decoction of roots of Godhapurna is also taken.

(k) Name of Disease: Khujali (Scabies)

Symptoms: Itching.

Therapy: To cure scabies, the Saharia take bath with hot water in which juice of crushed roots of Neem is mixed. After bath the patient is allowed to sit in sunlight for one or two hours. Leaves of Pila-Datura are also rubbed on the sites of irritation to cure scabies. At the time of sleeping, they apply Karanj oil over the entire body. Tari (toddy of Neem) is taken regularly for a month to cure the disease of Khujali. Turmeric (Haldi) solution is also given to the patients. Paste of root-powder of Tarwan or Tarawar and mustard oil is also applied on the irritated part of the skin. Mahua oil is massaged for curing Khujli. Leaves of Kandol are crushed into paste to be used as an ointment and applied on the skin. Roots and flowers of Kaner are used extensively to cure scabies.

(l) Name of Disease: Ulti (Vomiting)

Symptoms: Cold, heat and Anapacha (indigestion), dehydrated.

Therapy: Decoction of leaves of Sanjeevani is also taken. Mixture of Ajawaine and Kalanamak are given to the patient. The Siraka (vinegar) of Mahua, Jamun, lemon, etc., is served to the patient.

(m) Name of Disease: Dasta (Loose Motion)

Symptoms: Cold, heat, sun-stroke, eating rotten food.

Therapy: Gum of Babool is taken orally. Leaf- juice of Brangraj is taken orally with honey. Fresh juice of Khurao plant is given with sugar or gur (jaggery). The patient is given the leaves of Tulsi (basal plant) in Gur to eat, Tari of Khajur or Tar is given to the patient. Siraka of Jamun, Phuldhawai, Lemon, is served to the patient. Ripe fruit of Bel is also given to the patient.

(n) Name of Disease: Daanta Darda (Toothache)

Symptoms: Irresistable pain in the teeth.

Therapy: Roots of Datura are chewed or used like toothbrush to cure toothache. Twigs of Tarawar are used for brushing teeth. Mixture of Basil (Tulsi) leaves in hot water is given to the patient. Leaves of Bor are ground in warm water and gargled with, for curing toothache. Its branches are used for brushing teeth. Decoction of flowers and leaves of Siras is used for gargling. Juvenile leaves of Kair are chewed.

(o) Name of Disease: Pet Darda (Stomachache), Branch of Bor tree with gargling in hot water.

Symptoms: Irresistable pain in the stomachache.

Therapy: Crushed roots of Kanti are taken orally. Decoction of Cheela leaves, decoction of Adak-Datura roots and Arkmala roots is taken. The patient suffering from stomachache is given a mixture of Tulsi leaves, Gur (jaggery) and honey. Kalanamak and Ajawaine are also served to the patient. Decoction of Gawar-Patha leaves along with salt and lemon juice is also taken to cure stomachache.

(p) Name of Disease: Anakha Aana (Infection of eye)

Symptoms: Pain in the eyes, eyes become red, and difficulty in vision.

Therapy: Carbon collected by burning seed-oil of Pili-kanteli is applied on the eyes. Warmed leaves of Kesundo are put over the eyes to cure conjunctivitis. Decoction of Kumbat flowers is used to wash the eyes. Ground paste of Neem is applied on the eyes. Leaves of Babool dipped in warm water and put on the eyes of the patient. The Saharia cure the infection of eyes by using the milk of goat. They drop goat-milk in the eyes at the time of sleeping. The night soil of Girgit (lizard) is collected and dissolved in a little quantity of water. This solution is dropped in to the eyes at the time of sleeping.

(q) Name of Disease: Bawaseer (Piles)

Symptoms: Acute pain during defecation, discharge of anal blood during defecation, loss of body weight and weakness.

Therapy: Extract of roots of Bel is taken for about a week to cure piles. Powder of Akra flowers is taken with black pepper orally. Oily and spicy food materials and meat are prohibited. Tari (toddy of Neem) is given to the patient regularly.

(r) Name of Disease: Bhukha (Loss of Appetite)

Symptoms: The patient does not want to eat or drink anything, general weakness.

Therapy: Saharia have the knowledge that when a man starts losing his appetite, his health begins to deteriorate. If a man has good appetite, his health remains sound. Losing appetite is considered bad for the health. In order to increase the appetite, Saharia prepare a dry mixture of Amla, Hare and Bahera. This mixture is taken with honey or Gur. Leaves of Chawli are cooked and taken to cure loss of appetite.

(s) Name of Disease: Pathari (Stone)

Symptoms: The patient has difficulty in urinating, pain in the abdomen.

Therapy: Decoction of Bui flowers is taken for promoting discharge of urine. The Saharia try to cure Pathari by eating the pulse of Kurathai and the fruit of

Pathal Phorawa plant. The powder of tubers of Motha is taken with water for the treatment of urinary obstructions. The patient is advised to eat the flesh of Titir, Bater, etc. Decoction of Chawli roots is also taken.

(t) Name of Disease: Pilia (Jaundice)

Symptoms: Colour of urine becoming yellow, eye and the entire body becomes yellowish and growing weakness. The body appears bloodless. The patient runs slow temperature and experiences pain in upper abdomen.

Therapy: A great precaution is taken on the intake of diet. Oily and spicy food materials and meat are prohibited. The patient is allowed to take Khichari. Patient is given honey in the morning. Pulp of leaves of Rambans is taken to cure jaundice. The syrup of Bel fruits is taken. Root powder of Narkant is taken with water. Tulsi leaves and Gur are also served to the patient regularly. The leaves of Shisham plant is crushed and mixed in Matha (left after churning curd) and served to the patient regularly.

(u) Name of Disease: Dhata (Night fall)

Symptoms: Seminal discharge in the urine. The patient becomes lean and thin.

Therapy: A mixture of Amala, Hare and Bahera is prepared. The patient is advised to take this mixture regularly early in the morning with Gur. Chilly and sour food items are strictly tabooed. Seeds of Babul are soaked in cow-milk for two to three days, then the seeds are dried in shade, then the paste of seeds is taken with sugar. Fresh decoction of Khari-booti roots is taken orally by males.

(v) Name of Disease: Safed Paani (Leucorrohoea)

Symptoms: Discharge of whitish fluid, emission of foul smell from the genitals. The patient becomes lean and thin.

Therapy: A mixture of Amala, Hare and Bahera is prepared and given to the patient. Oil of Amla is applied on the genitals. The gum of Babool is dissolved in water and the solution is applied on the sex organs. Leaves of Babool are also taken with roots of Narkanto for the treatment of leucorrohoea. Paste of leaves of Dholi-moosli and extract of flowers of Semal is taken orally.

(w) Name of Disease: Banjha (Barrenness)

Symptoms: Among the Saharia, if a woman does not deliver a child after five or six years of marriage, she is believed to be suffering from barrenness. The Saharia believes that barrenness is caused due to irregular meeting menstruation periods.

Therapy: The ladies eat raw tubers of Hedulo to promote fertility and vitality. Fruits of Imli and Gur are also taken regularly. Decoction of Kiramar plant is taken. Fresh decoction of leaves of Khari-buti is given for regularising menstruation. When menstruation becomes regular, this removes the barrenness caused by irregular meeting of menstruation period. The extract of Kahua and Amaltas plant's bark is prepared and served to the patient suffering from barrenness. This regularises the flow of menstruation. When the flow of menstruation becomes normal, the chances of becoming pregnant grow.

(x) Name of Disease: Kabj (Constipation)

Symptoms: Loss of appetite, unable to defecate for two to three days.

Therapy: Tablets are prepared from the paste of roots of Kholi and one tablet is taken daily in empty stomach as a laxative to cure constipation. Fruit pulp of Amaltas is also taken. Extract of stem-bark of and leaves of Dhediya are taken.

Allopathic Treatment

In addition to the magico-religious, herbal and indigenous methods of treatment the Saharia also follow allopathic system of medicine. In the past the Saharia always used to consult traditional healer for treatment of sickness. But now, with the advancement, faith in supernatural concept is on decline and the Saharia have faith in allopathic doctors also. But the Saharia who live in interior areas mainly practice ethnomedical system. 61.3 per cent Saharia of Upreti and 58.7 per cent Saharia of Taleti have ever utilized the PHC services. It has been observed that Saharia (Taleti) have been taking treatment from PHC for various diseases such as tuberculosis (40.3%), skin diseases (3.8%), malaria (0.96%), asthma (1.9%), typhoid (0.96%), nightblindness (0.96%), leucorrhoea (0.96%), fits (0.96%) and accident (0.96%) etc. Saharia (Upreti) have been taking treatment from PHC for various diseases such as tuberculosis (37.8%), skin diseases (3.2%), asthma (1%) and accident (1%) etc (Tables 60 & 61). However it was observed that in most villages (75%) different treatments (spiritual, herbal and allopathic) go side by side and people have equal faith in all the treatments. They don't give preference to any particular method of treatment. It was told by some of the school teachers that Saharia are not able to purchase medicines due to poverty. Even if they could purchase they cannot continue them for a long period and this results in discontinuance of treatment after a short period. They also change doctors quite often during treatment. At an advanced stage of the disease many a times doctors also fail to treat it. Then the Saharia again have to resort to their traditional methods of treatment or continue to move from one doctor to another and spend a substantial amount of money on treatment. Continuity of sickness for a long time is not uncommon among the Saharia. The longer the period of sickness the higher is the expenses on treatment.

According to Doctors of PHC Shahabad the most common ailments for which the Saharia approach the PHC are skin diseases like scabies and eczema, loose motions, malaria, infection of upper respiratory tract, pneumonia, injuries, conjunctivities and ailments due to malnutrition. Few Saharia have more faith in private practitioner (Unqualified) and prefer to spend money on him rather than going to the government hospital or PHC. The doctors in the PHC Shahabad also told that the Saharia have more faith in injections than in tablets. They want immediate recovery and think it can be provided by an injection. A Saharia would take injections prescribed to him but has a callous attitude towards tablets. However in the interior villages it was found that when a person falls ill initially he consults a traditional healer, locally called 'Janteyar'. When the treatment of 'Janteyar' did not give any relief and the disease enters it's critical stage then the patient consult an allopathic doctor. The new generation is changing with the modern times. But the Saharia of older generation still adhere to ethnomedical practices of curing diseases. It is only when the 'Janteyar' fails to cure the patient and his condition becomes bad to worse then patient is taken to the allopathic doctor for treatment.

The services of qualified doctors are inaccessible to a majority of Saharia, especially in interior villages and where these are accessible the poor Saharia patients are neglected by the doctors. In such circumstances, they prefer to go directly to nearby private unqualified doctors where good treatment at low cost is available.

Saharia respondents were also interviewed for the time frame for treatment of illness, first approach for treatment, consultation with family members, visit to PHC, PHC service utilization, pattern for consultation of health services (traditional vs modern) which have been discussed below:

Table 41 presents that the Saharia of Taleti (18%) and Upreti (22%) don't go for treatment of illness immediately. 26 per cent Saharia of Taleti go within 1-10 days for treatment as compared to Saharia of Upreti (6%). The Saharia of both Taleti (42%) as well as Upreti (56%) who go for treatment between 11-20 days is higher because the Saharia of these villages (Beelkheradang, Matiakhad, Chaurakhadi, Pathari, Sandhri Khandasehrol, Ghesua, Hatri, Dhikwani and Mahodra) are interiorly located and also due to lack of transport facilities they are not able to go for treatment at Primary Health Centre (PHC) or Sub-Centre (SC). It was also observed that 14 per cent Saharia of Taleti and 16 per cent of Upreti who don't take any treatment were also living in the interior villages.

Table 41. Time Frame for Treatment.

S. No.	*After how many days you go for treatment of illness when you fall ill?*	*Saharia of Taleti*	*Saharia of Upreti*	*Saharia Total*
1	Don't go Immediately	27 (18%)	33 (22%)	60 (20%)
2	1-10 days	39 (26%)	09 (6%)	48 (16%)
3	11-20 days	63 (42%)	84 (56%)	147 (49%)
4	Don't take any treatment	21 (14%)	24 (16%)	45 (15%)
5	Total	150 (100%)	150 (100%)	300 (100%)

Table 42 presents the first approach for treatment. It has been observed that 47.3 per cent Saharia of Taleti and 28 per cent Saharia of Upreti first approach 'Janteyar' for treatment of illness. Only 22 per cent Saharia (Taleti) and 32.7 per cent Saharia (Upreti) go to PHC/Govt doctor for treatment of illness.

Table 43 presents the responsibility for consulting family members before taking any treatment. It has been observed that 78.7 per cent of Taleti and 65.3 per cent of Upreti Saharia consult their family members before taking any treatment.

Table 42. First Approach for Treatment

S. No.	*Whom you approach first when you fall ill?*	*Saharia of Taleti*	*Saharia of Upreti*	*Saharia Total*
1	'Janteyar' (Traditional Healer)	71 (47.3%)	42 (28%)	113 (37.7%)
2	PHC/Govt. Doctor	33 (22%)'	49 (32.7%)	82 (27.3%)
3	Private Doctor (Unqualified)	26 (17.3%)	38 (25.3%)	64 921.3%)
4	None	20 (13.4%)	21 (14%)	41 (13.7%)
5	Total	150 (100%)	150 (100%)	300 (100%)

Table 43. Consultation for Treatment.

S. No.	*Whether you consult your family member before taking any treatment?*	*Saharia of Taleti*	*Saharia of Upreti*	*Saharia Total*
1	Yes	118 (78.7%)	98 (65.3%)	216 (72%)
2	No	32 (21.3%)	52 (34.7%)	84 (28%)
3	Total	150 (100%)	150 (100%)	300 (100%)

From Tables (44, 45 & 46) it can be observed that 60 per cent of the Saharia have ever utilised the PHC services. But the Saharia of both Taleti (58%) and Upreti (64%) were not satisfied with the facilities available at the PHC. 67 per cent Saharia (Taleti & Upreti) said that free medicines are not available at PHC. 32 per cent Saharia of Upreti said that free medicines are available at PHC but stated that doctors give prescription on paper only and patients were not given any medicine from PHC. Under these circumstances the Saharia consult Private doctor

Table 44. Visit to PHC by Saharia.

S.No.	*Have you ever visited PHC?*	*Saharia of Taleti*	*Saharia of Upreti*	*Saharia Total*
1	Yes	88 (58.7%)	92 (61.3%)	180 (60%)
2	No	62 (41.3%)	58 (38.7%)	120 (40%)
3	Total	150 (100%)	150 (100%)	300 (100%)

Table 45. Satisfaction with the Services of PHC

S. No.	*Are you satisfied with the services of PHC?*	*Saharia of Taleti*	*Saharia of Upreti*	*Saharia Total*
1	Yes	63 (42%)	54 (36%)	117 (39%)
2	No	87 (58%)	96 (64%)	183 (61%)
3	Total	150 (100%)	150 (100%)	300 (100%)

(Unqualified) who gives proper treatment. The Private doctor charges fees for diagnosis and medicines. It may be mentioned here that only 58.7 per cent Saharia of Taleti and 61.3 per cent Saharia of Upreti have ever used the PHC services.

Table 46. Availability of Free Medicine at PHC

S. No.	*Is free medicine available at PHC ?*	*Saharia of Taleti*	*Saharia of Upreti*	*Saharia Total*
1	Yes	51 (34%)	48 (32%)	99 (33%)
2	No	99 (66%)	102 (68%)	201 (67%)
3	Total	150 (100%)	150 (100%)	300 (100%)

From Tables (47, 48 & 49) it can be observed that if the Saharia are not cured by 'Janteyar' then they consult PHC or Private doctor (Unqualified). This percentage was found to be higher among the Saharia of Upreti (74.7%). 38 per cent Saharia of Taleti said that they will not consult

Table 47. Pattern for Consultation of Health Services (Traditional vs Modern): Curing Practices by Janteyar or PHC/Private Doctors.

S. No.	*If not cured by 'Janteyar' whether you consult PHC/ Private Doctor (Unqualified)?*	*Saharia of Taleti*	*Saharia of Upreti*	*Saharia Total*
1	Yes	93 (62%)	112 (74.7%)	205 (68.3%)
2	No	57 (38%)	38 (25.3%)	95 (31.3%)
3	Total	150 (100%)	150 (100%)	300 (100%)

PHC or Private doctor if they are not cured by the 'Janteyar' and they will stop taking any treatment (Table 47). Mostly Saharia of Taleti (54.6%) and of Upreti (35.3%) told that if they are not cured by 'Janteyar' then they go to PHC/Private doctor after 10 to 20 days. It has been observed that mostly Saharia don't consult allopathic Doctor immediately after illness. Only 8 per cent Saharia of Taleti and 25.3 per cent of Upreti said that they consult PHC/Private doctor immediately if they are not cured by the 'Janteyar' (Table 48). When Saharia are not cured by the PHC or Private Doctor then they revert back to

Table 48. Pattern for Treatment of Modern Health Services: Time Framework.

S. No.	*If yes, after how many days you go to PHC/Private Doctor (Unqualified) for treatment ?*	*Saharia of Taleti*	*Saharia of Upreti*	*Saharia Total*
1	Immediately	12 (8%)	38 (25.3%)	50 (16.7%)
2	1-10 days	28 (18.7%)	42 (28%)	70 (23.3%)
3	11-20 days	82 (54.6%)	53 (35.3%)	135 (45%)
4	21-30 days	28 (18.7%)	17 (11.4%)	45 (155)
5	Total	150 (100%)	150 (100%)	300 (100%)

'Janteyar' for treatment. The percentage of Saharia of Taleti who revert back to 'Janteyar' was 58 per cent as compared to Upreti (27.3%) (Table 49). This percentage was higher in interior villages because Saharia of older generation adhere to old practices of curing diseases besides taking allopathic treatment also.

It has been found that out of a population of 5675 among Saharia of Taleti (8 villages) there were 1435 (25.3%) sickness cases whereas in Upreti (8 villages) the number of sickness cases were 816 (22.8 %) out of a population of 3583 as reported by the medical doctors at PHC. It is to be mentioned here that the data was collected from 104 sickness cases in Taleti and 95 sickness cases in Upreti.

Table 49. Pattern for Curing Practices of Modern Health Services vs Traditional System: Attitudes towards Traditional Healing Practices.

S. No.	*If you are not cured by PHC/ Private Doctor (Unqualified) whether you revert back to 'Janteyar' for treatment ?*	*Saharia of Taleti*	*Saharia of Upreti*	*Saharia Total*
1	Yes	87(58%)	41(27.3%)	128 (42.7%)
2	No	63(42%)	109(72.7%)	172 (57.3%)
3	Total	150 (100%)	150 (100%)	300 (100%)

Out of 104 sickness cases in Taleti, 9.6 per cent patients were taking herbal treatment, 29.8 per cent patients were taking treatment spiritual ('Janteyar') treatment, 52.9 per cent were consulting PHC and 7.7 per cent were not taking any treatment. Out of 95 sickness cases in Upreti, 10.5 per cent patients were taking herbal treatment, 24.2 per cent patients were taking spiritual treatment ('Janteyar'), 43.1 per cent were consulting PHC and 22.2 per cent were not taking any treatment (Tables 49 & 50).

From Tables 50 & 51 it can be stated that the incidence of tuberculosis (60.8%) is very high among the Saharia. This is mainly because of their low living standard, poor personal hygiene, smoking and drinking habits and other social/environmental conditions. The incidence of skin diseases like scabies ('Dad') and eczema ('Khujali') is also high (11.5 per cent in Taleti and 16.8 per cent in Upreti) due to unhygienic living conditions. Other diseases prevalent among Saharia are nightblindness ('Ratondi'), malaria ('Chotte Maharaj'), fits ('Dakini Lagna' or 'Bhoot Lagna' or 'Upri Hawa'), typhoid ('Bade Maharaj') diarrhoea ('Ulti-Dast'), stones ('Pathar'). During the fieldwork it was found that for snake bite cases Saharia always consulted 'Janteyar' because they think that doctors are helpless in curing snake bites and they can do anything.The Saharia of Taleti (59.6%) and Upreti

Table 50. Name of Diseases and Types of Treatment Followed by Saharia of Taleti.

S. No.	*Name of Disease*	*Number of Cases*	*Types of Treatment*			
			Herbal	*Spiritual 'Janteyar'*	*Allopathic*	*No Treatment*
1	Tuberculosis (TB)	62 (59.6%)	4 (3.8%)	12(11.5%)	42 (40.3%)	4 (3.8%)
2	Skin diseases (Scabies, Eczema)	12 (11.5%)	2 (1.9%)	4 (3.8%)	4 (3.8%)	2 (1.9%)
3	Malaria/Fever	4 (3.8%)	—	3 (2.8%)	1(0.96%)	—
4	Asthma	3 (2.8%)	—	1 (0.96%)	2 (1.9%)	—
5	Typhoid	5 (4.8%)	1 (0.96%)	3 (2.8%)	1 (0.96%)	—
6	Snake bite	2 (1.9%)	—	2 (1.9%)	—	—
7	Nightblindness	4 (3.8%)	1 (0.96%)	2 (1.9%)	1 (0.96%)	—
8	Leucorhoea	4 (3.8%)	1 (0.96%)	1 (0.96%)	1 (0.96%)	1 (0.96%)
9	Fits	3 (2.8%)	—	2 (1.9%)	1 (0.96%)	—
10	Stones	2 (1.9%)	—	—	1(0.96%)	1(0.96%)
11	Diarrhoea (cough- cold headache & vomiting)	2 (1.9%)	1(0.96%)	1(0.96%)	—	—
12	Accident	1(0.96%)	—	—	1(0.96%)	—
	Total	104 (100%)	10 (9.6%)	31(29.8%)	55 (52.9%)	8 (7.7%)

Table 51. Name of Diseases and Types of Treatment Followed by Saharia of Upreti.

S. No.	Name of Disease	Number of Cases	Types of Treatment			No Treatment
			Herbal	Spiritual 'Janteyar'	Allopathic	
1	Tuberculosis (TB)	59 (62.1%)	6 (6.3%)	7 (7.4%)	36 (37.9%)	10 (10.5%)
2	Skin diseases (Scabies, Eczema)	16 (16.8%)	—	4 (4.2%)	3 (3.2%)	9 (9.5%)
3	Malaria/Fever	3 (3.2%)	—	3 (3.2%)	—	—
4	Asthma	1 (1%)	—	—	1(1%)	—
5	Typhoid	3 (3.2%)	—	3 (3.2%)	—	—
6	Snake bite	1(1%)	—	1(1%)	—	—
7	Nightblindness	2 (2.1%)	2 (2.1%)	—	—	—
8	Leucorrhoea	1 (1%)	—	—	—	1 (1%)
9	Fits	3 (3.2%)	—	3 (3.2%)	—	—
10	Stones	1(1%)	—	—	—	1(1%)
11	Diarrhoea (cough-cold, headache & vomiting)	4 (4.2%)	2 (2.1%)	2 (2.1%)	—	—
12	Accident	1(1%)	—	—	1(1%)	—
	Total	95 (100%)	10(10.5%)	23 (24.2%)	41 (43.1%)	21(22.2%)

(62.1%) who were suffering from TB (age ranges from 30-55 years) were found attending TB Control Camp which is held on 23rd of every month at Shahabad. It shows that the new generation is changing with modern times.

The Saharia have identified some diseases like fits, snakebite, typhoid, diarrhoea, TB, skin diseases, malaria/ fever etc., for which they mostly consult 'Janteyar'. If the 'Janteyar' fails to cure the patient only then they go to PHC for treatment.

However in case of accident they consult an allopathic doctor for getting treatment at the PHC because they think that 'Janteyar' cannot cure it.

Case Studies with reference to Ethnomedical System of Treatment

The following case studies will reflect the interaction of traditional and modern health care system emphasizing on socio-cultural perspective.

Case study I: Kailash, a resident of Khandasehrol, a female aged 24 years, is suffering from 'Upri Hawa' (Fits). According to her husband, she was taken to 'Janteyar' for treatment but she did not recover from illness. For the last three days she was screaming, dancing, shouting and singing. Now she is taken to Baran for treatment. According to Kailash's husband there are cases of 'Bhoot Lagna' or 'Upri Hawa' (Fits) in which the person starts singing, dancing and shouting. Then he/she is taken to 'Janteyar' one day before 'Poornamasi' or a day before 'Amavasya' (of Chaudas) then 'Janteyar' gives him/her the treatment. The 'Janteyar' gives 'Jhada' by taking bird feathers or plant twigs. The 'Janteyar' takes some maize grains in his hand chants some mantra and then goes in trance. His body starts shivering and then he tells the cause of the ailment. He tells the name of the deity which is angry with the sick person and tells the person to made offerings to that particular deity. Only then the sick person will be cured. The 'Janteyar' is given ghee, coconut and sugar. The 'Janteyar' doesn't take any money because if he takes money his power will be finished. The people in interior villages are more superstitious. They will leave the settled colony if four to five deaths occur in the Saharana.

Case study II: Kanwar Lal, aged 44 years, a resident of village Beelkhera Dang, told that "I was having fever and initially I thought that there was no need of consulting an allopathic doctor because my condition was not serious. I consulted a 'Janteyar' for treatment. But when my condition started deteriorating slowly then I rushed to PHC at Shahabad and I was given prescription on paper only and was not given any medicine from PHC. Then I stopped going to PHC at Shahabad and now I am going to Sipri in M.P. for getting treatment at a hospital". Kanwarlal further mentioned that "at PHC the doctor give prescription on paper only. The patients were not given any medicine and the villagers are not happy with the services available at PHC. There are cases of fits prevalent in the village. For fits a 'Janteyar' is consulted. The 'Janteyar' goes, in trance. His body starts shivering rapidly and the 'devi' or 'devta' comes to him then he explains the cause of illness and tells the patient to made offering to a particular deity. The deities which comes to him are mainly Thakur, Jind, Khains, Bheran, Kali and Bijasun Mata. If the deity does not come to him then it means the patient will die after some time".

Case study III: Shravan, aged 35, a resident of village Pathari, narrates " when a child dies while taking milk or during breast feeding, it is believed to be caused by a ghost known as 'Kaccha Machan'. The dead body of the child is kept in a place, digged and then covered with mud. For avoiding such death the 'Janteyar' ties a black thread around neck and forearm of the children ('Band lagana'). It protects the child from evil-eye".

Case Study IV: Man Singh, aged 55 years, a resident of Khusalpura village, narrates "I am having infection in the eyes. I am following the indigenous treatment for this disease and I found the treatment very effective. He mentioned that the paste of leaves of 'Bamoor' tree if applied on the eyes or the milk of goat is put as eye drops for one week then eye infection is cured".

Case study V: Chaita, aged 32 years, a resident of Shahpur is suffering from tuberculosis for the last three years. She describes "I went to PHC Shahabad for medical check up three years back. But I did not followed the entire course. After one year my condition became serious then I consulted a private doctor (Unqualified) but there was no improvement in my condition." She further mentioned "there are no medicines at PHC. A TB control camp is held at Shahabad on 23rd of every month. Initially I used to attend TB control camp for three to four months but after that I stopped going to attend the TB camp. Then I went to hospital at Baran. The facilities are good at Baran. There is an X-ray machine also. I was admitted at Baran for three months and now I am perfectly all right. Medical facilities are not good at PHC Shahabad. There is an X-ray machine but it is not working and doctors also sometimes scold the patients and then they stop going to PHCs. Then the patients consult private doctors for treatment but it requires money and they cannot afford to consult Private Doctor every time. Then they go to bigger hospitals at Baran or Kota. The life of poor people is like hell. She cried *'Hamari koi sunwai nahi hai'* (No body listens to us)".

Case study VI: Raghubir, aged 37 years, is residing in the village Faredua. He is suffering from night blindness (Ratondi). He narrates "Initially I consulted a 'Janteyar' for treatment. But there was no improvement in my condition then I went to Shahabad for treatment where I was told to go to Kota to get treatment in the big hospital. In all this process I have already spend a lot of money." He further said "when 'Janteyar' fails to cure the disease then we consult private doctor or go to PHC and even if the doctors at PHC fails to bring any improvement in the condition of the sick persons then we go to Baran or Kota for treatment".

Case study VII: Dharamjeet, aged 35 years, lives in the village Dhikwani. His wife is suffering from T.B. According to him "there is no proper arrangement for treatment in villages. If any one has any trouble all of sudden, he or she is taken to Shahabad PHC in a bullockcart where many a times doctors refuse to see the patient. Then we are examined by the compounder who asks for money and does not give medicines from the hospital. Dharamjeet also reported that Government doctors do not give proper treatment. No one pays attention to us. Only private doctors treat us properly".

Case Study VIII: Saku Bai, aged 42 years, an anganwadi worker at Shahabad said "Since the Saharia are poor, in case of sickness, they first try herbal remedies and watch for two to three days. If there are not cured than they go to Primary Health Centre (PHC) or Sub-Centre (SC) for treatment. The Saharia are not satisfied with the facilities available at the PHC. The doctor gives prescription on paper only and the patients were not given any medicines from PHC. Sometimes the doctors also refuse to see the patients and then the Saharia consult private doctors who pay attention on them. But they have to spend large amount of money on this process. The Saharia are not afraid of immunization. When PHC staff visit their colonies then they get their children immunized." It was found that all the children below one year were immunized.

Case Study IX: Urmi, aged 31 years, resident of village Gangapura narrates "In the past I always used to consult 'Janteyar' (faith healer) for treatment of sickness. But now, I have more faith in allopathic doctors. However, in a few cases I take different treatments (spiritual, herbal and allopathic) also side by side".

Case Study X: The faith of women in 'Janteyar' seems to be shaking as Mangla, aged 42 years, resident of Devri village said "I believed that 'Janteyar' (traditional healer) will cure my child. 'Janteyar' also assured me that the child would get well soon. On his assurance my faith in him continued and I left everything to him. The faith and reliance on him led to death of my child".

Case Study XI: Neeru, age 38 years, a resident of Sandhri village says "Last month my son aged 14 years was suffering from fever with shivering. My husband contacted 'Janteyar' and his treatment continued for three days but the treatment was not found effective, then we took him to PHC Shahabad. The doctor was not available and the compounder on duty gave some tablets and charged Rs. 25/-. We returned home and continued the treatment for four to five days but there was no improvement, than we approached to a Private Doctor living in the nearby village. He gave injections daily and my son was cured within five days. The doctor charged Rs 50/- for the treatment". She further said "The new generation is changing with the modern times, though the Saharia of older generation still adhere to old practices of curing diseases and only when the 'Janteyar' (faith healer) fails to cure the patient and his condition becomes bad to worse then he is taken to the allopathic doctor for treatment".

It was also observed during the study that though most of the Saharia have visited atleast once the PHC, but they did not express any satisfaction regarding the services provided at these centres and they would not like to visit them again for their treatment. They complained that doctor's do not pay proper attention and generally neglect them. Many times, doctor's even refuse to see the patients. The Saharia, therefore, prefer to consult private doctor's who pay more attention to them.

Case Study XII: Ramli, aged 24 years, a resident of Chaurakhadi village says "Government doctor's do not give good treatment. Government doctor's do not check us properly and even refuse to see the patients. No one pays attention to us. So, in most of the cases, we go to private doctor's who treat us properly."

Case Study XIII: Ghasi, aged 55 years, a resident of village Hatri said "I am suffering from TB for the last two years. First I went to Shahabad to attend the TB control camp but my condition did not improved. The doctor at the PHC also told me to stop coming to PHC. Then I went to Guna (MP) for treatment at a private hospital but there was no improvement in the condition. I spent a lot of money. Now I am not taking any treatment". He further mentioned "for minor ailments such as low fever, cough, cold, the tribals first go to 'Janteyar' who sees the 'nadi' or 'nabj' (pulse) of the patient and gives herbal medicine ('Jungli Dawai' or 'Jadi Booti'). However if the symptoms continue and the condition of the patient doesn't improve than the patient would go to PHC and if the symptoms continue further than the patient would go to private doctor. If again there is not any improvement in the condition of the patient then he would go to hospital at Baran for treatment".

Case Study XIV: Kajri, aged 32 years, a resident of village Mahodara narrates "I have four children, I did not take any injection or tablets during my pregnancies because, as per my mother-in-law, these are harmful for the child in the womb and even may lead to the death of the child before delivery" She further said "The doctor's in the PHC ask for money when we go to PHC for treatment. We would like to go to PHC but due to unsympathetic attitude of doctors and compounder at PHC we have stopped going to the PHC. The compounder asks for Rs. 25/-Rs. 50/- for injection and Rs.10/- Rs.20/- for a tablet. The Doctor also asks sometimes upto Rs 200/-. Under such circumstances we would like to consult private practitioner who has a sympathetic attitude towards the tribals."

Case Study XV: Raghuvar, aged 38 years, a resident of Chaurakhadi said "One sister (nurse) and one compounder should be there at the Sub-Centre. For the last two years there is no sister and compounder in the Sub-Centre. A dispensary should be made in the village. For the last seven years the village has Kaccha road. Pucca road should be made because when a person is seriously ill than it becomes very difficult to take him to Shahabad or Baran or Shipri (MP) for treatment at PHC or hospital. When there is any emergency than we have to call a private doctor who sometimes ask for Rs. 200-300/- also. The road that goes from Devri to Chaurakhadi to Bheel Khera Dang is a Pucca road on paper only. Infact it is a Kaccha road".

Negligence by the doctor's is not the only factor responsible responsible for the low utilization of health services. The misconceptions entertained by Saharia regarding the quality of medicines and faith in injection as the only effective remedy are also responsible for not utilising health services. The Saharia have more faith in injections than in tablets. Syrups and tablets are not much effective according to them. If they are not prescribed injection by the Government doctors, they feel that they are not being provided with good treatment as mentioned in the following case study.

Case Study XVI: Dali, aged 34 years, living in the village Ghesua says "I have delivered five children but I never visited any doctor or hospital for any problem related to pregnancy. At the time of delivery of the second son, premature bleeding started, I took some medicines from the 'Janteyar' (traditional healer) and I was cured. Last year, one of my neighbour's wife had some problems during delivery, the doctor was called but of no help and the child died. All these things depend on the God who decides to whom he wants to cure". She further said ``The PHC staff does not give good treatment. Either they give water like syrup or few tablets, but not injections. The PHC staff turns us out after giving few tablets. So, we go to private doctor, pay him fee and get treated." She also complained about the quality of food given at the Aanganwadi and mentioned "The food given at Aanganwadi is not of good quality. But due to hunger and poverty children eat this food and fall ill."

Focus Group Discussion (FGD) – Saharia of Taleti

Two FGD were carried out among the Saharia of Taleti. Eight male individuals were selected in each FGD, who belonged to middle age group (30-40 years). All the participants were illiterate. Their villages (Pathari & Sandhri) were not connected with pucca road. Components taken into consideration for FGD were; concept of illness, health seeking behaviour, attitude towards traditional healer and felt need of tribal population with regard to health services.

Concept of Illness

1. 75 per cent participants from Pathari and 87.5 per cent participants from Sandhri reported that whenever they felt physically uncomfortable they considered themselves sick and remaining participants from both villages reported that when they could not carry out daily routine work they considered themselves sick.

Health seeking Behaviour

2. All the participants (100%) reported that there was no doctor near their village whom they could

approach for seeking treatment. All the participants (100%) revealed that two important factors which influence treatment were firstly severity of ailment and secondly proximity of the doctor. All the respondents (100%) revealed that for minor illness they preferred traditional healer who resides in their village. But if the illness got severe then they go to Sipri (M.P.) 75 per cent respondents from Pathari and all the respondents from Sandhri mentioned that when they go to PHC which is at a distance of 30 Kms from their villages, the doctors did not give them much attention and medicine were also not available at PHC so they don't prefer to go to PHC.

3. 62.5 per cent respondents from Pathari and 75 per cent from Sandhri reported that whenever any one fell sick, after waiting for one or two days, the patient was taken to private practitioners who reside near their village. For minor ailments traditional healer ('Janteyar') is consulted.

4. All the respondents (100%) from both the villages revealed that private doctors were preferred for prompt attention and quick relief. All the respondents reported that private doctor gave more attention to patients. 50 per cent respondents from Pathari and 50 per cent from Sandhari also reported that private doctors used stethoscope, checked pulse of the patient and also listened to patient problems.

5. 62.5 per cent respondents from Pathari and 75 per cent from Sandhri also revealed that when the condition of patient becomes very serious they were taken to either block PHC or Hospital at Sipri (M.P.). In fact during serious condition private doctors themselves referred the patient to these places.

6. As revealed by 87.5 per cent participants from Pathari and all the respondents from Sandhari said that the transportation system was very poor and patients found it difficult to travel by crowded buses.

7. 62.5 per cent respondents from Pathari and per cent 75 per cent from Sandhri revealed that doctors at PHC did not spend much time with patient's problem and also said that medicines were not available in BPHC. After waiting for long time, doctor wrote the prescription and medicine was supposed to be purchased from the chemist shop at Shahabad. Doctors were also in hurry as a result it became difficult for the patients to understand his advice.

Attitude Towards Traditional Healer

8. 87.5 per cent respondents from Pathari and 87.5 per cent from Sandhri reported that traditional healers ('Janteyar) were preferred, in order to eradicate the evil power from person whenever child cries continuously or if the person was sick for a long time, In such cases the help of 'Janteyar' was sought. These respondents also revealed that the help of 'Janteyar' was needed when a person is bitten by snake.

Focus Group Discussion (FGD) – Saharia of Upreti

Two FGD were carried out among the Saharia of Upreti. Eight male individuals were selected in each FGD who belonged to middle age group (30-40 years). In the village Hatri, out of eight individuals, four had completed their primary education, two had completed secondary school education and remaining two were illiterate in the village Ghesua. PHC is at a distance of 20 km from the village Hatri.

In village Ghesua 8 males aged (30-35) were selected for interviews. Noneof them was literate. Same components were taken into consideration for Saharia of Upreti. The PHC is at a distance of 9 Km from Ghesua.

Concept of Sickness

1. 37.5 per cent participants from Ghesua and 50 per cent participants from Hatri reported that whenever they felt physically uncomfortable they considered themselves sick. Rest of the participants from both the villages reported they considered themselves sick only when they couldn't get up from bed unable to comply with daily routine work.
2. 65 per cent participants from Ghesua and 50 per cent participants from Hatri reported that cold and cough, bodyache, low temperature fever, diarrhoea and skin diseases were not considered unusual, in other words, individual showing any of these symptoms were not considered as sick. Since they were very poor, they couldn't afford to continue the treatment for long time and also were unable to buy medicines for every small cause.
3. 75 per cent participants from Ghesua and all the participants (100%) from Hatri reported that when there was nothing to eat at home, expenditure on health became a secondary factor except when the person was very sick.

Preference for Providers

4. 62.5 per cent participants from Ghesua and 37.5 per cent participants from Hatri reported that they used to go to Primary Health Centre (PHC) whenever they were sick.

5. All the participants from both the villages felt that choice of specialist depended upon seriousness of illness. Cost for the treatment was an important factor to decide whether they would seek treatment or not was felt by 75 per cent participants from Ghesua and 65 per cent participants from Hatri.
6. 62.5 per cent participants from Ghesua complained that compounders and nurses were very rude and they treated poor with great degree of indifference. They also reported that medicine was not available at PHC or Sub-Centre and patients were asked to buy medicine from chemist shop at Shahabad. However, the rest 37.5 per cent participants reported that medicines were available at PHC but the nurse or compounder don't gives medicines to them.

Attitude towards Traditional Healer

7. 87.5 per cent participants of Ghesua reported that the help of traditional healer ('Janteyar') was sought in cases like snake bite and fits etc. However all the participants (100%) of Hatri reported that cases of snake bite and fits were cured by 'Janteyar'.

MATERNAL AND CHILD HEALTH CARE

It has been observed that 88 per cent Saharia do not give any special diet to the new born except the breast feed. The solid food is given normally after one year or so in most cases. The implications of this observation are born out by prevalent high rates of malnutrition (Sukha Bimari) among the Saharia Children. A pregnant woman is not any nutritious food. She takes the routine diet during the period of pregnancy. Sometimes milk is given. The pregnant woman works in field also. The Saharia women know that intake of milk, dalia, green leafy vegetables is good for the growth of foetus. But due to poverty they cannot afford to eat what they should eat during pregnancy. Delivery conditions in interior villages were found to be unhealthy. There is no ('Bhangan') local sweepress in the interior villages. The ('Bhangan') is called to cut the umbilical cord and charges Rs.200/- to Rs.300/- on birth of a female child and Rs.300/- to Rs.400/- on birth of a male child. When a delivery is to be conducted, she is called from Shahabad and which sometimes takes two to three days. Till then, the mother and the newly born baby, remain unattended. They are not allowed to eat or drink anything until the umbilical cord is cut and washing is done. This affects the mother and child's health and leads to dire consequences. A large number of children, therefore die before completing one year, have high IMR (85.9). After delivery, the mother is given jaggery, mahua water to clean internal organs and remove impurities from the body if any. During the maternity period, the mother is fed on 'dalia' or 'halua' prepared from jaggery and jowar or wheat flour. They avoid taking ghee for three days as it may cause tetanus to the child. Oil and urad dal are not taken for one month as these are considered as heavy things. Sour substances including curd are taken after two months as these can cause swelling and pain. Thus, restrictions are many but prescriptions advising special diets are few. Pregnant and lactating mothers do not get any special nutrition. Consumption of ghee and milk is rare, even after the completion of the period of restrictions, due to poverty, Thus, a Saharia mother hardly gets nutritious diet during pre-natal and post-natal days. Similar is the position of the newly born infants, as they have to depend on the milk of malnourished mothers. During the first year, children are not given any supplementary foods. Only a few children get additional milk. Due to weakness, children are not able to eat bread. During the second year also children are able to eat bread in small quantity. Hence, a majority of children like their mothers do not get sufficient food and both remain malnourished.

Table 52 and Fig 7 shows the distribution of Saharia women in accordance with availing ANC checkup and number of visits for ANC check up. It has been observed that 64.7 per cent Saharia women of Upreti and 38.7 per cent of Saharia women of Taleti have utilized ante natal care (ANC) services. 61.9 per cent of the Saharia women of Upreti and 65.5 per cent Saharia women of Taleti visited the health centre only once. 25.7 per cent women (Upreti) and 24.1 per cent (Taleti) paid two visits to the health centre. 12.4 per cent women (Upreti) and 10.4 per cent (Taleti) have paid three visits to health centre.

Table 52. Distribution of Saharia Women in Accordance with Availing ANC Checkup and Number of Visits for ANC Check up

Saharia				*Number of visits for ANC Check-up*				
Area	*Yes*	*No*	*Total*	*1st*	*2nd*	*3rd*	*4th*	*Total*
Upreti	97 (64.7)	53 (35.3)	150 (100.0)	60 (61.9)	25 (25.7)	12 (12.4)	–	97 (100.0)
Taleti	58 (38.7)	92 (61.3)	150 (100.0)	38 (65.5)	14 (24.1)	6 (10.4)	–	58 (100.0)

Figures in parenthesis denotes percentage

Table 53 presents the distribution of women utilizing ANC services vis-à-vis distance of villages from Health Centre in Upreti. It has been observed that 64.7 per cent of Saharia (Upreti) visited health centre for ANC services whereas 35.3 per cent did not visit for the same. Among the Saharia women of Upreti where health centre is located within the village 50.5 per cent women have utilized ANC services at the health centre, when health centre is located within 5 km from the place of their residence 12.5 per cent women have utilized the ANC services, 9.2 per cent women have utilized ANC services

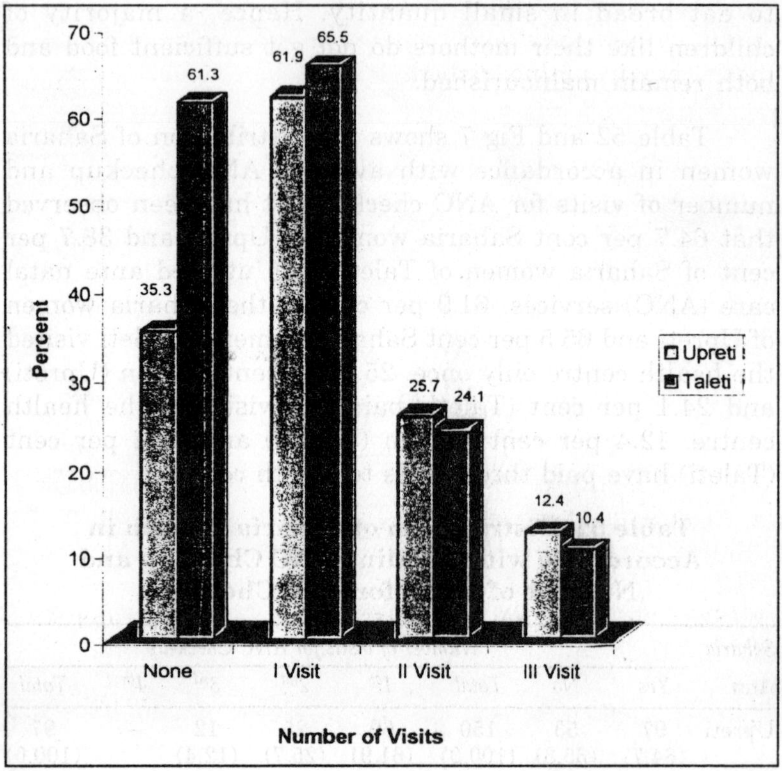

Fig. 7. Number of Antenatal Visits.

of health centre situated at a distance of 5-9 km from their residence, 27.8 per cent women have utilized ANC services of health centre situated at a distance of more than 10 km from their residence. Out of the sample who did not visit the health centre, it has been observed that 28.3 per cent Saharia women did not visit the health centre which is located within their villages. 54.7 per cent of Saharia women did not visit health centre because the health centre was 10+ km away from their village as well as lack of transport facilities.

Table 54 shows distribution of women utilising ANC services vis-à-vis distance of villages from Health Centre

Table 53. Distribution of Women Utilizing ANC Services vis-à-vis Distance of Villages from Health Centre (Upreti).

Area	Distance 9 Km	Total Number of Villages	No. of Households Covered	Visited Health Centre	Not Visited Health Centre
Upr.	Within Village	3 (37.5)	64 (42.7)	49 (50.5)	15 (28.3)
	< 5	1 (12.5)	16 (10.7)	12 (12.5)	4 (7.5)
	5-9	1 (12.5)	14 (9.3)	9 (9.2)	5 (9.5)
	10+	3 (37.5)	56 (37.3)	27 (27.8)	29 (54.7)
	Total	8 (100.0)	150 (100.0)	97 (64.7)	53 (35.3)

figures in parenthesis indicate percentage

(Taleti). Among the Saharia women of Taleti where health centre is located within the village 44.9 per cent women have utilized ANC services at the health centre, when

Table 54. Distribution of Women Utilising ANC Services Vis-à-vis Distance of Villages from Health Centre (Taleti).

Area	Distance	Total Number of Villages	No. of Households Covered	Visited Health Centre	Not Visited Health Centre
Tal.	Within village	2 (25.0)	46 (30.7)	26 (44.9)	20 (21.7)
	< 5	1 (12.5)	14 (9.3)	4 (6.9)	10 (10.9)
	5-9	–	–	–	–
	10+	5 (62.5)	90 (60.0)	28 (48.2)	62 (67.4)
	Total	8 (100.0)	150 (100.0)	58 (38.7)	92 (67.3)

figures in parenthesis indicate percentage

health centre is located within 5 km from the place of their residence 6.9 per cent women have utilized the ANC services, 48.2 per cent women have utilized ANC services of health centre situated at a distacnce of more than 10 km from their residence.

Reproductive Health Problems of the Females and Folk Medicine

The pregnancy (*garabha*) is welcomed in the Saharia society because it provides a chance to a husband to prove his musculinity and a female to prove her feminity. It provides a chance to the couple to continue the progeny. The Saharia ladies were reluctant to talk about such sensitive issues. Only after establishing rapport it has been made possible to collect information pertaining to reproductive health problems of the females. The problems faced by the pregnant women are weakness (*kamjori*), swelling (*soojan*), prolapse (*bacchedani kamjor hona*), irregular bleeding and excessive bleeding during menstruation (*gano khoon chale*). Folk medicine is used for care of pregnant women. Solution of unrefined sugar in water (*gur*) and turmeric powder *(haldi ka doodh)* is given to pregnant ladies. The TBA (*dai*) helps the pregnant ladies in successful delivery (*prasauti*). She also does some massage with Mahua beer on the waist and thighs of the pregnant woman to remove the delivery pain. The Mahua beer is also spread on the parts nearby reproductive organs. Paste of fresh leaves of *Morpankhi* is taken with fresh cow-milk (one glass) for a week to regularise menses period. The ash of shade dry plant *Morpankhi* (about 2-5gms) mixed in fresh cow-milk (one glass) or with honey and jaggery (*gur*) is given to a lady for a fortnight twice or thrice a day after menses period who wish to have an issue. Powder of *Semal* flowers with honey is given to ladies suffering from excessive bleeding during menstruation. Juice of *Khari-buti* is given to ladies for regularising menstruation. Flowers of *Safed-jhojhru* or *Bhurbhura* are used against erruptions and suppression of menstruation. When the flow of menstruation becomes normal, the chances of becoming pregnant grow. Among Saharia the pregnant and nursing mother face the situation of anemia due to lack of haemoglobin content in the blood of the woman. The Saharia do not have knowledge about haemoglobin but they call anemia as lack of blood. In order to cure anemia indigenous medicine is used. Paste of nutritious roots, shoots, flowers and fruits available in and around the forest is prepared after boiling, crushing and cooking them in Mahua beer. From this paste mixture, small round pieces are made and taken as medicine. The extract of barks of Pipal, Bar, Augusta, Amaltas, etc. is prepared and served to the patient. Infusion of *Nawari* plant is given as a growth tonic. Extract is prepared from *Narkanto* roots and sugar and one teaspoonful is taken daily as a health tonic for pregnant women. Decoction of *Punarva* roots is also taken to increase the blood content. Saharia also use some indigenous medicine in order to increase the milk of the mother for proper feeding of the child. The mother is advised to take the milk of goat. She is advised to eat green leafy vegetables.

For termination of illegal pregnancies (pregnancies resulting from pre-marital and extramarital relationships) help of traditional healer (*'Janteyar'*) is sought. A mixture of Mahua flowers, giger, black pepper and jaggery (*gur*) is given. It induces bleeding leading to abortion. It has been observed that in 90 per cent cases Saharia women consult *'Janteyar'* for abortion. The *'Janteyar'* mostly terminate pregnancies resulting from pre-marital and extra-marital relationships. They mostly terminate pregnancies up to three months. Bark of *Rohan* is boiled in water with jaggery (*gur*) and given to the pregnant lady for causing abortion. However if there are any further complications the Saharia then consult private (unqualified) doctors. In one or two discussions mention has been made of ANM also for consultation regarding reproductive problems. On being asked why don't they go to PHC or Sub-centre for treatment of these problems the Saharia women mentioned that due to absence of lady doctors at PHC they don't go to PHC. The ladies feel hesitant to talk to male doctors regarding reproductive problems. The doctors in the PHC Shahabad mentioned that due to poverty the Saharia women indulge in unsafe sex with the strangers and suffer from these problems.

In 98 per cent villages the Saharia women are unaware about HIV or AIDS. In 2 per cent villages the ladies mentioned having heard about HIV or AIDS. But they don't know how it is transmitted. The community leaders and NGO workers especially CECOEDECON mentioned that leucorrhoea (*Safed Paani*) is very common among the Saharia women. Use of indigenous medicine, poverty, ignorance, lack of knowledge and superstitious beliefs are the reasons mentioned by the community leaders for it. According to them the Saharia women regard these problems as part of their lives.

The Saharia believe that barrenness is caused due to the presence of evil spirits like Bhoot, Preta, Dain, etc. In such case, 'Janteyar' (traditional healer) is consulted who tries to drive away the evil spirit. When such efforts become vain, it is believed that some powerful Dain has cast black magic on her.

Focus Group Discussion (FGD) – Saharia of Upreti

Two focus group interviews were carried out among the Saharia of Upreti with regard to Maternal and Child Health Care (MCH). Interviews were carried out in two Upreti villages (Khandasehrol and Dikhwani). FGD was carried out among a group of six women in each village.

The interview was carried out among six ever married women in the age group of 25-30 years from each of the villages. None of the women from these villages completed their primary school education. Components taken into consideration for FGD were; awareness about ANC check-ups; factors restraining women from utilising ANC services, consumption of food during pregnancy and child immunization.

Awareness about ANC check up

1. 59 per cent of the participants related ANC with the consumption of iron and folic acid tablets and tetanus toxoid injections. 41 per cent of the participants expressed their unawareness.
2. 51 per cent of the participants felt that pregnancy was a physiological event and did not require any medical care.

Factors restraining women from utilizing ANC check-up

3. 50 per cent participants from these villages felt that their mother-in-laws do not allow them for medical check-up. The mother-in-law feared that such check-up might harm the child. It was also reported that their mother-in-laws rely more on elderly ladies and traditional dais. They also expressed that they could not rely on anganwadi workers and health workers for ANC check-up, as they are young and have little knowledge and skill regarding birth of a child.
4. Communication and transport were the major problems expressed by 40 per cent respondents. 50 per cent participants reported that it is not possible for pregnant women to go for regular check-ups. The bullock-carts were often the sole mode of conveyance between the peripheral village and the nearest health facility. It took long

time and hence difficult for women to travel by these means. Private vehicle became too expensive for regular check-up during pregnancy. 60 per cent participants reported that the state Government buses were operating from Baran to Sipri but these buses are over crowded.

5. 28 per cent participants reported that no health personnel visited their village for distribution of iron tablets or for immunizing pregnant women.

Consumption of Food during Pregnancy

6. All the participants reported that usual food was consumed during pregnancy. No special care is taken of pregnant women among the Saharia. She takes the routine diet during the entire period of pregnancy. Milk is given only if they have milching cattle.

Child Immunization

7. 45 per cent of the participants from these villages of Upreti could tell utility of each specific vaccination and number of vaccinations. However 60 per cent participants from the villages revealed that polio vaccine saves the child from physical handicap.

Focus Group Dicussion (FGD) – Saharia of Taleti

Two FGD were carried out among two villages (Beelkhera Dang and Faredua) of Taleti. FGD was carried out among a group of six women in each of the two villages. None of the women were literate components taken into consideration for FGD were same that of the Saharia of Upreti.

1. 58 per cent of the participants expressed their unawareness regarding ANC check-up. Only 42 per cent were aware of it.

Necessity of ANC check-up

2. 66 per cent of the participants felt that there was no need for ANC check-up during pregnancy since it was believed that pregnancy was a natural process.

Factors restraining women-from utilizing ANC check-up

3. 84 per cent participants reported that since the Block PHC was 15-20 km. away from the village, it was not possible for them to go for regular check-up. It was also felt by respondents that they were not aware of the time, when health worker visited their village, since they used to go out for work early in the morning. To avail such services, would result in loss of their daily wage. 68 per cent participants also told that someone had to accompany pregnant women, this would result in loss of wages for both of them. BPHC was always over-crowded and women had to spend 2-3 hours at BPHC for different check-ups from one person to another and it became very tiring.

Food consumed during pregnancy

4. All the participants from both these villages maintained that usual food was consumed during pregnancy. No special care is taken of a pregnant woman among the Saharia. Saharia woman takes the routine diet during entire period of pregnancy.

5. 82 per cent women felt that instead of such preventive measures if food or money were provided to them to meet their needs, it would have been more beneficial.

Child Immunization

6. 32 per cent women participants from these villages could name the different vaccination,

number of doses or their utility. However 42 per cent women felt that although they were aware that immunization was good but they were not aware about exact time and the number of doses of vaccination.

Family Planning

Table 55 reveals the awareness regarding family planning methods and sources among the Saharia women of Taleti and Upreti. It can be observed from the Table 55 and Figure 8 that 82 per cent women of Taleti and 92 per cent of Upreti were knowledgeable about sterilization as a method of FP. 2.7 per cent women of Taleti and 2.0 per cent of Upreti were aware about oral pill, only 2 per cent women of Taleti were aware about IUD, 10.6 per cent women of Taleti and 6.0 per cent of Upreti were aware about traditional method of FP and 2.7 per cent women of Taleti were aware about Multiple choice (more than one method of FP). Thus, the knowledge about FP method among the Saharia women of Taleti and Upreti is mainly restricted to terminal method of FP.

In order to access the influence of mass media FP, respondents were asked whether they had seen any material on FP. It can be observed from Table 55 and Figure 9 that 58.7 per cent Saharia women from Taleti and 52.0 per cent Saharia women from Upreti had never seen any material on FP. Further 25.3 per cent women from Taleti and 32.0 per cent from Upreti had seen film show on FP. 6.7 per cent women (Taleti) and 2.0 per cent (Upreti) had seen TV/listen to radio on FP. 9.3 per cent women (Taleti) and 8.0 per cent (Upreti) had seen puppet show on FP. Only 4 per cent and 2 per cent women of Upreti had seen pamphlets as well as posters on FP whereas not a single woman from Taleti had seen the same.

Similarly, in order to know about the sources of FP messages respondents were asked whether anybody had

Table 55. Awareness Regarding Family Planning Methods and Sources of Spreading Family Planning Messages

Statement	*Saharia*		
	Taleti	*Upreti*	*Total*
I. Awareness of FP Methods			
1. Sterilization	82.0%	92.0%	87.0%
2. Oral Pill	2.7%	2.0%	2.35%
3. IUD	2.0%	–	1.0%
4. Traditional Method	10.6%	6.0%	8.3%
5. Multiple choice	2.7%	–	–
Total	100.0%	100.0%	100.0%
II. Influence of Mass Media			
Nil	58.7%	52.0%	55.35%
Film show	25.3%	32.0%	28.65%
Puppet show	9.3%	8.0%	8.65%
Pamphlets	–	4.0%	2.0%
Posters	–	2.0%	1.0%
TV/Radio	6.7%	2.0%	4.35%
Total	100.0%	100.0%	100.0%
III. Anybody spoken about FP			
PHC staff/Anganwadi workers	52.0%	54.0%	53.0%
Neighbours/Friends	24.0%	20.0%	22.0%
Pvt. Practioners	16.0%	14.0%	15.0%
No body	4.0%	6.0%	5.0%
Village Dai	4.0%	6.0%	5.0%
Total	100.0%	100.0%	

Note: Figures in parenthesis denotes percentage

spoken to them about FP. It was found that 52.0 per cent respondents from Taleti and 54.0 per cent respondents from Upreti were reported to have come to know about FP from PHC/Anganwadi workers. Another 24.0 per cent

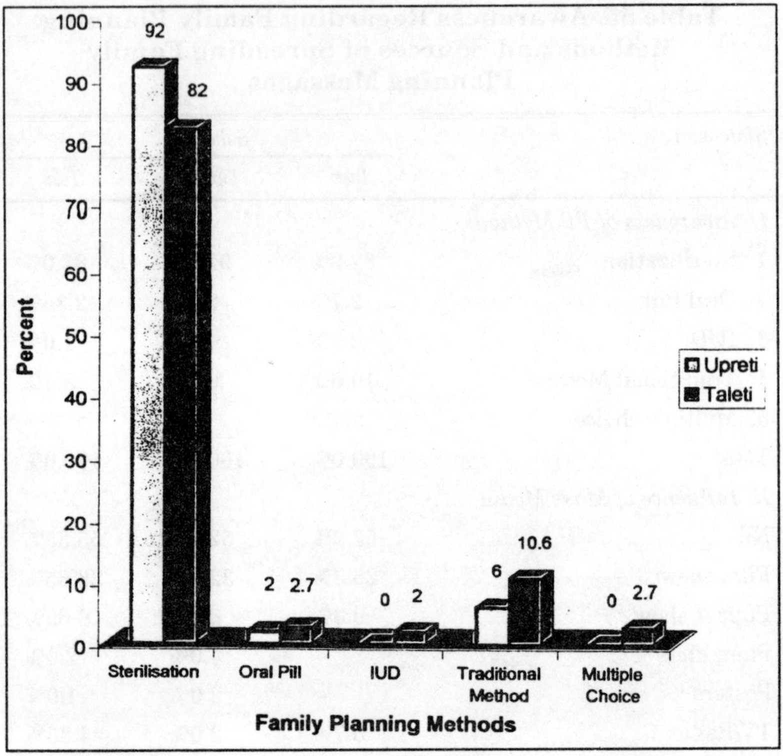

Fig. 8. Awareness About Family Planninig Methods.

(Taleti) and 20.0 per cent (Upreti) have come to know about FP through interpersonal communication (friends/neighbours), 16.0 per cent women of Taleti and 14.0 per cent of Upreti have come to know about FP from private practitioners. 4.0 per cent women (Taleti) and 6.0 per cent (Upreti) have come to know about FP from village Dai only (Table 55 and Figure 10).

Table 56 and Figure 11 reveals the attitude of Saharia of Taleti and Upreti towards family planning. The attitude of respondents towards restriction on the number of children in the family has been found. It was observed that 64 per cent Saharia of Taleti and 86 per cent Saharia of Upreti wanted to restrict the number of children.

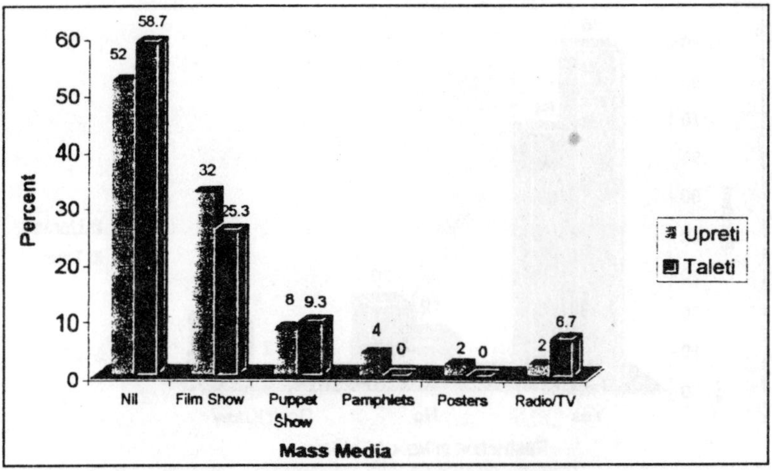

Fig. 9. Influence of Mass Media.

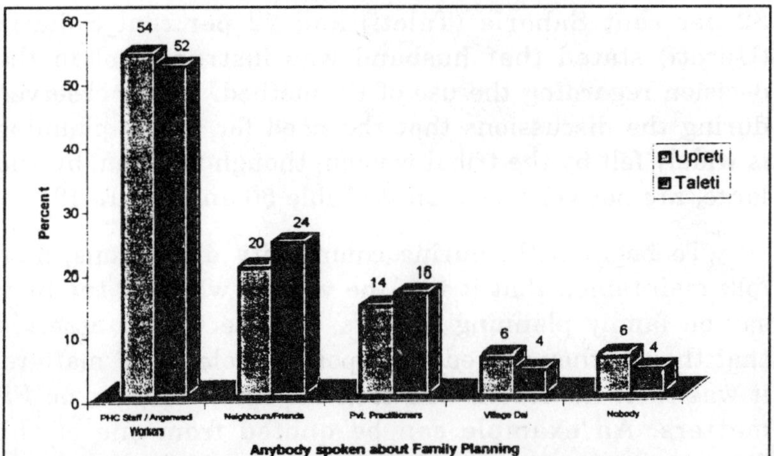

Fig. 10. Person spoken about Family Planning.

As far as decision making regarding the kind of FP method was concerned it was revealed by 16 per cent Saharia (Taleti) and 12 per cent Saharia (Upreti) that it was the mother-in-law who were decision maker. However 8 per cent Saharia (Taleti) and 16 per cent Saharia (Upreti) revealed that the decision for adopting FP method was the joint decision of both husband and wife. Another

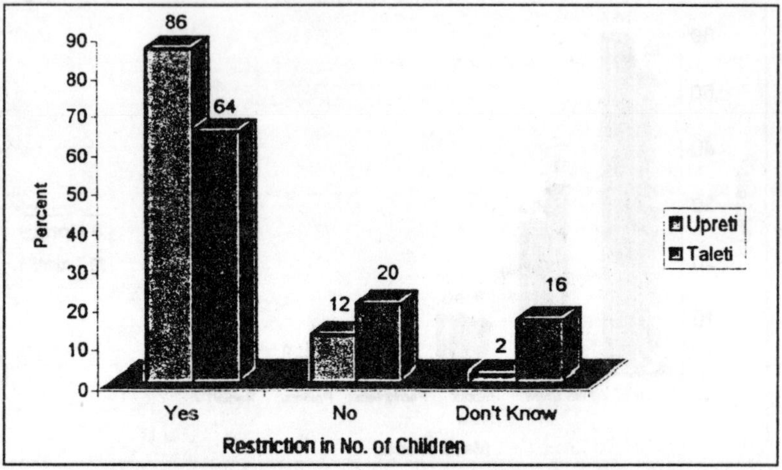

Fig. 11. Attitude towards Family Planning.

32 per cent Saharia (Taleti) and 22 per cent Saharia (Upreti) stated that husband was instrumental in the decision regarding the use of FP method. It was observed during the discussions that the need for family planning is widely felt by the tribal women, though the men, by and large, are not very keen on it (Table 56 and Figure 12).

To begin with, during community discussions, men folk maintained that it was the women who had the final say on family planning matters; but the women asserted that though they played an important role in FP matters, it was their husband's who took the final decisions on FP matters. An example can be quoted from one of the discussions. In case a woman has two or three daughters and wishes to undergo tubectomy, she cannot do so without the consent of her husband. In such cases, the husband's mostly are keen on having atleast one male child, and they continue to reproduce till they get the desired number of male children. In case a woman is reluctant to co-operate, her husband threatens to acquire another woman, who could bear him sons. In such cases, according to the women, they have to, yield to their husband's desire.

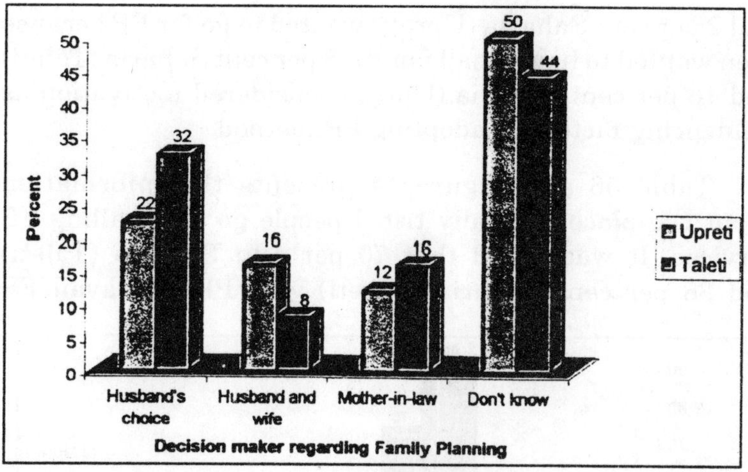

Fig. 12. Decision making regarding Family Planning.

Factors, which influence a person for adopting FP methods has been shown in Table 56 and Figure 13. The responses of those who have ever used any FP measures and who have not used were taken into consideration. Poor economy was the reason for 32 per cent Saharia (Taleti) and 62 per cent Saharia (Upreti). 28 per cent Saharia (Taleti)

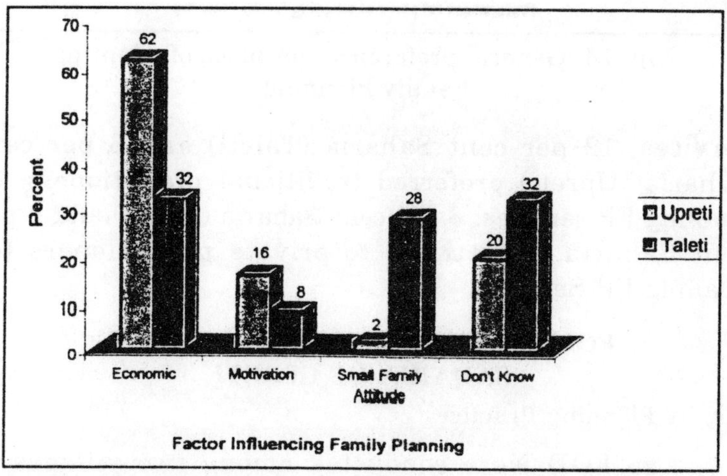

Fig. 13. Factor Influencing Family Planning Adoption.

and 2 per cent Saharia (Upreti) wanted to go for FP because then wanted to have small family, 8 per cent Saharia (Taleti) and 16 per cent Saharia (Upreti) considered motivation as influencing factor for adopting FP method.

Table 56 and Figure 14 presents the information about the place normally tribal people go for availing FP services. It was found that 70 per cent Saharia (Taleti) and 88 per cent Saharia (Upreti) visit PHC to avail FP

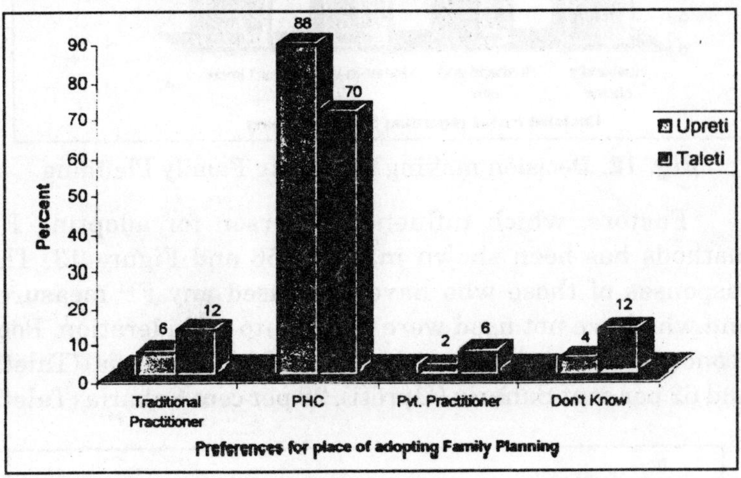

Fig. 14. General preferences for place of adopting Family Planning.

services. 12 per cent Saharia (Taleti) and 6 per cent Saharia (Upreti) preferred traditional practitioners for availing FP services. 6 per cent Saharia (Taleti) and 2 per cent Saharia (Upreti) go to private practitioners for availing FP Services.

FOCUS GROUP DISCUSSION (FGD) – SAHARIA OF UPRETI

Family Planning Practices

Two FGD were conducted among two villages of Upreti (Samrania and Mahodara). Each FGD consisted of

Table 56. Attitude Towards Family Planning Among Saharia of Taleti and Upreti

Statement	*Saharia*		
	Taleti	*Upreti*	*Total*
I. Restriction in No. of Children			
Yes	64.0%	86.0%	75.0%
No20.0%	12.0%	16.0%	
Don't know	16.0%	2.0%	9.0%
Total	100.0%	100.0%	100.0%
II. Decision Maker Regarding FP			
Husband's Choice	32.0%	22.0%	27.0%
Husband and Wife	8.0%	16.0%	12.0%
Mother-in-law	16.0%	12.0%	14.0%
Don't Know	44.0%	50.0%	47.0%
Total	100.0%	100.0%	100.0%
III. Factors Influencing FP Adoption			
Economic	32.0%	62.0%	47.0%
Motivation	8.0%	16.0%	12.0%
Small Family Attitude	28.0%	20.0%	24.0%
Don't Know	32.0%	20.0%	26.0%
Total	100.0%	100.0%	100.0%
IV. General Preferences for Place of Adopting FP			
Traditional Practitioner	12.0%	6.0%	9.0%
PHC	70.0%	88.0%	79.0%
Pvt. Practitioner	6.0%	2.0%	2.0%
Don't Know	12.0%	4.0%	8.0%
Total	100.0%	100.0%	100.0%

Note: Figures in parenthesis denotes percentage

six married women in the age group of 25-35 years. Two women were having education up to middle school. Components taken into consideration in FGD were awareness of FP, sources of awareness, attitude towards FP, centre for adopting FP methods and felt need in FP.

Awareness of FP

1. Sterilization was most widely known method among respondents. Knowledge regarding reversible methods was limited. 48 per cent of participants were familiar with the terms IUD, Copper T and Oral Pills.

2. 52 per cent participants were aware of the person who provided traditional medicine for birth control, but could not give much information regarding the kind of medicine given.

Sources of Awareness

3. NGO's like CECODECON (Centre for Community Economics and Development Consultants Society), LOK JUMBISH PARIYOJNA, SANKALP were the major sources for spreading FP information among both the Saharia of Taleti and Upreti. PHC workers have also played a role in spreading FP messages. Inter-personal Communication has also played an important role in spreading FP information.

4. Mass media had limited impact in spreading FP messages as expressed by respondents. However 34 per cent respondents had seen pamphlets and posters during FP camps.

5. 62 per cent respondents from these villages reported that health workers did not give any or much information about reversible methods of FP. Much stress was given on sterilization. The respondents felt that the health workers were interested merely in knowing whether the couple had undergone sterlization or not.

6. Information regarding reversible methods provided by health workers was not satisfactory as revealed by 82 per cent of respondents. They also expressed

their own fear with the use of reversible methods, participants revealed health personnel deliberately did not tell the side-effects of FP methods, fearing that nobody would then opt for FP.

Attitude towards FP

7. 89 per cent of respondents reported that they were not in favour of FP due to following reasons:
   (a) They felt that after adopting the FP the mothers become very weak and therefore, were not able to look after the family properly.
   (b) They feared that if something ill happened to their existing children, they would not be able to produce any more children in future, after the operation.
8. 61 per cent respondents favoured FP, as they felt that due to their poor economic condition they would not be able to feed a large family.

Centres for adopting FP methods

9. 98 per cent respondents preferred PHC as the place for adopting FP.

Focus Group Discussion (FGD) Saharia of Taleti

Two FGD were carried out among two villages of Taleti (Shahabad and Chaurakhadi). Each FGD consisted of six married women in the age group of 25-35 years. All the participants were illiterate. Same components were taken into consideration.

1. All the respondents were aware of sterilization. However only 5 per cent participants were aware regarding Oral pills, copper T, IUD.

Attitude towards FP

2. 64 per cent participants favoured FP mainly due to their poor living condition and their inability to support large family.

Place for Adopting FP

3. 88 per cent respondents preferred PHC as the place for adopting FP.

ROLE OF HEALTH SERVICE PROVIDERS

Allopathic Health Providers

Allopathic health providers practicing in the Upreti and Taleti include doctors at PHC and private practitioners. Both Government and private health providers were interviewed for their roles regarding creating awareness and acceptability of health care delivery system among tribal people. They were also interviewed to identify constraints, needs that could provide meaningful suggestions for a effective delivery of health care system.

Role of Medical Officers and Doctors at PHC

Medical Officers at Block Primary Health Centre (BPHC) and Doctors at PHC were medical graduates and none of them had received any specialized training, like medical post graduate (MD/MS) etc. Medical Officer is reponsible for undertaking promotive, preventive and curative work in an integrated manner.

1. Medical officer (MO) is given responsibility to look after the dispensary, out patient department and in-patient wards and supervises the duties to ensure smooth functioning of PHC. MO also received cases referred to him by health worker and trained dais.

2. Further Medical Officer and Doctors at PHC provide guidance to health assistants, health workers in the treatment of minor ailments.

3. Medical Officer and Doctors at PHC have also to ensure that health assistant, health workers and trained dais are fully conversant with the health

programmes to be implemented in the villages where they are working.

4. Medical Officer will actively involve health workers and trained dais in the effective implementation of MCH, FP and nutrition programmes.

5. Medical Officer and Doctors at PHC also organize camp meetings, health education talks, and display of posters with the assistant and help of health worker.

6. Medical Officer has to regularly up date the inventory and stock registers of equipment supplied to him, prepare indent for drugs, instrument etc. sufficiently in advance and submit them to appropriate health authorities. He has to scrutinize the programmes of his staff and suggested changes, if necessary to suit the priority of work.

7. Medical Officer PHC has to take corrective measures in the control of communicable diseases and proper improvement of sanitation in the villages.

8. Medical Officer PHC is directly responsible for controlling any out break of epidemic that might occur in the area.

9. Medical Officers and doctors at PHC organizes and participate in village health committee, village panchayat meeting to discuss health programmes with the community members, to identify their needs and to seek their cooperation.

10. Medical Officers and doctors at PHC holds a monthly meeting with health workers for evaluating the progress of work and suggests steps to be taken for effective implementation of various programmes.

View of Medical Officers With Regard to Tribal Health Need

Medical Officers at PHC are mainly treating cases of malaria, diarrhoea, scabies, eczema, cough, fever. But it was observed that special stress was given on child immunization and family planning methods. Limited laboratory facilities are available at PHC. Although vehicle facility is being provided at BPHC, it is used only for bringing medicine from district headquarters or Kota. Most of the drugs prescribed to the patient had to be purchased by the patients from the chemist shop. However in the interior villages there is no chemist shop.

The providers felt that chief reason for poor health situation among the Saharia were lack of awareness, illiteracy, poverty, low socio-economic condition and bad sanitary conditions. Providers felt that most common diseases among Saharia were Tuberculosis, Scabies and Malaria. Providers also felt that poor means of communication between villages and health centre led to under utilization of health services. They also felt that due to high price of drugs and poor economic condition tribal people could not continue their treatment for long and stopped taking treatment as soon as they felt little comfortable without completing the full course of medicine. As a result their health condition did not improve. They approached health centre only when the condition of patients became worse. They blamed the traditional healer for under utilization of health centre by tribals. They felt that since these healers who reside in tribal villages, were better known among tribals. Tribals approach these healers more often as compared to PHC. Since these healers were not formally trained in allopathic medicine, often prescribed wrong medicine. It was also reported that in order to recover quickly, in some cases tribals observed to have taken overdose of antibiotic prescribed. Block Supervisor PHC Shahabad reported that BPHC was ill equipped to meet the needs of thousands of patients. It was running short of medicine and manpower. Health providers strongly felt that in order to improve the health condition among tribals there was a need for increasing literacy among tribals. Health providers also suggested that by improving the communication channel between providers and tribal community better delivery of healthcare could be ensured. Table 57 represents the view of MO with regard to tribal health.

Role of Health Workers (Male Health Workers and Female Health Workers)

At the grassroots level the health workers were the persons providing preventive and curative help to villagers. They had different sets of responsibilities for MCH, family planning, immunization and nutrition. In-depth interviews were taken from both male and female health workers regarding their functioning, constraint they faced and their views about the health needs of the community.

All male health workers were either matriculate or class 12th passed. Besides this they have been given training for their present job. For the present study six male health workers and six female health workers were interviewed. All the health workers told that there were seven Sub-Centres in the Block PHC Shahabad. Each health worker of Sub-Centre was expected to cover 5000 population, but in case of tribal population they were supposed to cover 3000 population. All health workers told that they have visited each village once in a month.

1. Health workers (both male and female) were supposed to identify cases of notifiable diseases i.e. cholera, poliomyelitis. They were supposed to identify symptomatic cases of TB, presenting symptoms like continued fever or prolonged cough or spitting of blood etc which they come across during their home visits and notify PHC Doctors and Medical Officers about them.

Table 57. View of Government Medical Officers With Regard to Tribal Health

Statement	*Block Medical Officer (BMO)*	*Medical officer (MO)*
Qualification	MBBS	MBBS
Views regarding Tribal Health	Poor	Poor
Reason for poor Health	Lack of awareness, poverty, illiteracy	Highly negligent, Bad sanitary condition.
Services Provided	Curative as well as preventive (MCH and Family Planning)	Curative Services, like treatment for malaria, scabies, TB
Prevalent Diseases	Tuberculosis,Asthma, Scabies, Malaria	RespiratoryTractInfection (RTI), Scabies, TB, Diarrhoea
Supply of Medicine	Shortage of medicines in the PHC, X-ray machine is not working.	Shortage of medicines in the PHC.
Constraints	Manpower was less. Due to poor economic condition tribals stop taking medicines after initial disappearance of symptom.	Ignorance of tribals also led to diseases, Presence of traditional healer also led to under-utilisation of PHC services.
Health Need of Tribals	There was a need to increase awareness among tribals. In interior villages pucca road should be made.	Need to increase literacy among tribals.

2. Duty of Health workers (both male and female) also includes chlorinating drinking water source including well at regular interval.

3. Both male and female Health workers were supposed to make arrangement for immunization once in a month at Sub-Centre and have to give prior information to villagers about the immunization camp.

4. Health workers (both male and female) have to spread the message of family planning in the village and motivate the villagers for adopting family planning measures individually as well in groups.

5. As far as maternal and child health (MCH) is concerned male health workers refer abnormal pregnancy cases i.e. women with medical and gynaecological prolems to female healh workers. Male health workers were also immunizing pregnant women with tetanus toxoid. Male health workers also have to record about births and deaths occurring in their area, general informaion about each village and submit fortnightly report to the supervisor at the PHC.

6. Health workers (both male and female) also provide treatment for minor ailments, ORS packets to diarrhoeal cases, first-aid for accident and refer cases beyond their competence to PHC.

7. Female health workers were given some additional responsibility in maternal and child health like registering and providing care to pregnant women throughout the period of pregnancy, referring abnormal pregnancy and cases with medical and gynaecological problems to PHC.

8. Female health workers also refer pregnant women to PHC for ANC check-up besides providing them wih iron and folic acid tablets. They also supervise deliveries conducted by dais.

9. Female health workers refer cases of difficult labour and new born with abnormalities to PHC. Male and female healthworkers have been given duties to educate mother individually or in groups for better family health, including MCH, family planning, nutrition, immunization, control of communicable diseases, personal and environmental hygiene and care of minor ailments.

10. Both male and female health workers accompany women to family camp and assist the Doctor.

Views of Health Workers with Regard to Tribal Health

Health workers also felt that the health status of tribals were poor. For this they blamed tribals. They felt tribals were highly negligent and illiterate because of this they did not utilise health services properly. Besides this their poor unhygeinic living condition, drinking and smoking habits were also responsible for their poor health. Health workers expressed that it was very difficult to convince the tribals for utilising preventive services. The health workers felt that utilisation of ANC was poor among tribal women. Though they supposed to make four visits for ANC but it was told that it was not so. Health workers told that if tribal women immunized their child, most of them did not complete the full course of immunisation. Health workers reported that there was tremendous work pressure on them. The villages covered by them were scattered and sometimes when they visit the household and the mothers were not available. It was also reported that due to shortage of staff they had to do additional work besides their normal duties. Female health workers reported that during rainy season, it usually becomes very difficult for them to cover all the villages of their sub-centre as the villages were cut-off from pucca roads. Health workers felt that illiteracy and unawareness were major hurdles for increasing their level of consciousness towards better healthcare. Table 58 gives the views of health workers with regard to tribal health.

ROLE OF TRADITIONAL HEALTH PROVIDERS

Role of TBAs

TBAs (Dais) advise the pregnant women about the intake of nutritious diet. They also tell the pregnant ladies to eat green leafy vegetables, milk. They also give some massage on the waist and thighs of pregnant ladies. The TBAs conduct the delivery however they refer the complicated cases to the ANM or to the big hospital at Baran or Kota. TBAs want to be integrated with the PHC or Sub-centre.The Saharia ladies are satisfied with the services of the ANM.However they mentioned that a lady doctor (gynaecologist) would be there in the PHC who could help them if there are any complications in pregnancy or labour pains or any other gynaecological problems.

Role of Traditional Healer ('Janteyar')

It has been observed that the same person acts as a traditional healer ('Janteyar'). Saharia go to them for treatment of diseases because they belong to Saharia community and they belong to Saharia community and they don't ask for money. They tie amulets ('Tabij') around the neck and forearm of the patient to drive away the evil-spirits. The 'Janteyar'also gives spiritual ash ('Bhabuti') to the patients. The 'Janteyar'also performs 'Jhara' with the twigs of *Akra* plant to ward off the evil eye. He also gives herbal medicines for various diseases. The 'Janteyar' also wants to be integrated with the Government hospitals

Table 58. View of Health Workers With Regard to Tribal Health

Statement	*Male health workers (MHW)*	*Female Health Worker (FHW)*
Educational qualification present job.	Matriculate besides training for present job.	Matriculates besides training for
View with regard to tribal health	Poor	Poor
Duty	At grass root level health workers were responsible for providing preventive and curative services.	Servicing the tribals with preventive and curative services
Constraints	Health workers reported that there was tremendous work pressure on them due to shortage of staff they have to do additional duties besides their normal duty. It is not possible to go to the villagers in the interior during rainy season.	It is difficult for them to cover all the villages of their sub-centre because the villages were cut off from the pucca road.
Health Need of Tribals	Awareness should be increased among tribals.	Awareness should be increased.

and health centres. They also seek some monetary incentives.

View of Traditional Healer ('Janteyar') with Regard to Tribal Health

Information was obtained from two traditional healers ('Janteyar') with regard to tribal health. Their view regarding tribal health, as reported by them is poor due to poor economic condition of tribals. Traditional healers ('Janteyar's) provided treatment for diseases like fits, skin diseases, malaria, tuberculosis, fever, treatment of those persons who were bitten by snake or poisonous creature and those suffering from other minor ailments. The duration of treating any person depends on the type of ailment the person was suffering from. Both the traditional healers informed that they could treat any type of disease, provided the patient came in time and obeyed their instructions. They usually charged the patients in kinds, whatever they could offer. But patient's family had to make arrangement for material needed for performing rights and rituals. Traditional healers reported that they did not have any professional contact with government health personnel. They felt that tribals preferred them, as far as they know their language and also reside in the same area of tribals. They were also concerned about financial status of patients.

Role of Private Medical Practitioners (PMPs)

When the Saharia don't get medicines from the PHC then they have to spend large amount of money on it. Under such circumstances the Saharia go to Private Medical Practitioners (PMPs). During fieldwork three Private Medical Practitioners (PMPs) were also interviewed. One of them was qualified (Ayurvedic system of medicine) and the other two were not qualified. They were called ('Bengali Doctor') in the village. They mentioned that thet have worked with qualified doctors for 2-3 years for experience and they have done MBBS from (Patna) Bihar. It has been observed that the Saharia go to these practitioners for treatment of diseases such as diarhoea, malaria, vomiting, fever, urino-genital disorders etc. The Saharia go to PMPs because they have a sympathetic attitude towards them. Thet listen to the problems of the patient. If the patient is unable to pay then they sometimes defer the payment to a later date. They also offer cheaper treatment and the duration of treatment offered by PMPs is shorter and effective. The private practitioners see around 70-80 cases around every month. They refer serious or complicated cases to big hospital at Baran.

View of Private Practitioners with Regard to Tribal Health

The PMPs mentioned that unsympathetic attitude of doctors and the para-medical staff at the PHC or Sub-Centre is one of the reasons responsible for dissatisfaction of Saharia with the services of PHC or Sub-Centre. In some interior villages the ANM charges upto Rs 250/- to Rs 300/- for supervising a delivery. The PMPs put emphasis on the development of roads and transports in the interior villages. More medicines should be made available at the health centres. There is a need of chemist shops also in the interior villages. There is a need of lady doctor for treatment of gynaecological disorders. The PMPs also wants to be trained in latest medicines and medical equipments. Private practitioners practicing in the area reported that many suspected TB patients were coming to them for seeking treatment. Besides, a number of cases of dysentry, cough, typhoid, cholera and skin infections etc., were being treated by them regularly. Among the children diseases such as *ulti* (vomiting), *dasta* (loose motions), *sardi-khansi* (cough and cold), and Respiratory Tract Infection (RTI) were reported to be very common. Private practitioners mentioned that the Saharia use indigenous and magico-religious methods also for treatment of various diseases. The private practitioners are unqualified.

The Saharia go to private practitioners because they are easily available and approachable and these private practitioners listen to their problems and also give proper attention to them. They respect the tribal culture. They refer complicated cases to the hospital at Baran. However the doctors at PHC sometimes scolds the patients and then they stop going to PHC. The private practitioners mentioned that uhygienic living conditions, non-availability of safe drinking water supply, lack of awareness, illiteracy and lack of medicines at PHC are factors contributing to the dismal health condition of the Saharia. Regular supply of medicines is required at the PHC. They said that health education is required among the Saharia. Their is a need of economic upliftment also among these people.

IMPACT OF HEALTH PROGRAMMES ON SAHARIA

In the Shahabad and Kishanganj tehsils, various programmes related to health and medical care have been running in the Primary Health Centres, Sub-Centres and dispensaries. These programmes are: Primary Healthcare, Rural Primary Healthcare Infrastructure, Health Manpower, Control of Communicable Diseases, National Malaria Eradication Programme, Mutipurpose Camps (for ENT diseases, treatment for Handicapped persons, Dental Check-up), National Tuberculosis Programme, National Leprosy Eradication Programme, Control of Non-Communicable Diseases, National Blindness Control Programme, Maternal and Reproductive Health Programme etc. The State and Central Government run some of these programmes.

It may be stated that some of these programmes have been running on the papers and the Saharia were not even aware about these programmes especially in the interior villages. They felt that whenever they contacted the PHC/SC/Dispensary/Health Centres they could not get sufficient information about these programmes. Saharia have been facing lot of problems about their medical treatment and healthcare institutions.

It was observed that Tuberculosis Control Camp is organised on 19th of every month at Kelwara and 23rd of every month at Shahabad. Many Saharia who are suffering from TB attend this camp. Saharia have been given 'Chainit Parivar' cards when they feel ill they can use this cards and get medicines from the PHC. But when interviewed many Saharia told that when they go to PHC the sister (nurse) or compounder ask Rs.2/- per tablet for cold and headache and if they don't give money to them then the compounder or nurse scolds them and don't give medicines. The Saharia are not afraid of immunization. When PHC Staff visit their villages then they get their children immunized. It was observed that all the children below one year were found immunized at the time of fieldwork.

The 'Nasbandi' (Tubectomy) camp is organised on 2nd and 4th Friday of every month at PHC, Shahabad. Table 59 presents the monthwise 'Tubectomy' cases among Saharia from April '99' to Jan. 2000 in PHC Shahabad.

It has been observed that maximum percentage of cases of family planning are done in November to January because Saharia females are not busy as they don't have any work to do for economic reasons. They are free and climate also suits them for undergoing operation. During summer the females are busy in various enonomic activities. In the interior villages the family planning is not much successful when compared to villages in the Upreti because means of communication are not there, absence of ANM due to vacant post and illiteracy. Only

Table 59. The monthwise 'Tubectomy' Cases Among Saharia from April '99' to Jan. 2000 in PHC Shahabad

S.No.	*Month*	*No.of Tubectomy Cases*
1	April 1999	7 (1.17%)
2	May 1999	1 (0.17%)
3	June 1999	3 (0.5%)
4	July 1999	8 (1.33%)
5	August 1999	NIL
6	September 1999	5 (0.83%)
7	October 1999	8 (1.33%)
8	November 1999	68 (11.33%)
9	December 1999	433 (72.17%)
10	Jan. 2000 (upto 7th Jan.)	67 (11.17%)
	Total	600 (100%)

women have been found adopting family planning (FP) measures. Males don't adopt FP measures because they think they will become physically weak and will not be able to work.

SAHARIA DEVELOPMENT PROJECT (SDP)

SDP was launched in 1977 under the chairmanship of District Collector. There are a number of schemes through which efforts are being made for the socio-economic upliftment of the Saharia under SDP. The project is being financed mainly through central assistance (94.33%) and 5.67 per cent from state plan. The various heads of the project include agriculture, irrigation, rehabilitation, rural housing, electrification, animal husbandary, forest development, medical and public health, drinking water, training, education, cooperation, free legal advice, road construction, other development programmes and administration. Implementation of SDP is the sole responsibility of Saharia Development Committee. The district collector heads the committee.

Members of the parliament and legislative assembly from the area, Zila Pramukh, two representatives of Saharia Community and related district level officers are members of the committee. Project officer of the Saharia Development Project is the Additional Divisional Magistrate (A.D.M.).

Funds have been provided under the SDP for free distribution of medicines to Saharia, establishment of Health Subcentres, organization of medical camps, family welfare programme and upgradation of PHC Shahabad into twenty beded hospital, along with residential quarters for the staff. However it has been observed that the implementation of medical and health schemes has been very slow. In the PHC Shahabad medicines are not available. X-ray machine is there but it is not working.

Voluntary Organisations

In addition several voluntary organisations like ADIM JATI SEWAK SANGH, CECOEDECON (Centre For Community Economics and Development Consultants Society), ASEFA and SANKALP are also working for socio-economic upliftment of Saharia. Adimjati Sewak Sangh was established on 14th November 1957. Mr. Badan Singh Verma is the Area organiser of the Adimjati Sewak Sangh. Badan Singh Verma came to Shahabad on 2nd August 1959 with an intention to work for the upliftment of the Saharia. The Sangh also created awareness about education, medical and health facilities, safe drinking water, family planning etc. Sangh has opened a hostel at Shahabad for education of Saharia children.

SANKALP is a voluntary organisation of youths, basically of Udaipur. They established a centre in village Mamoni. They organised camps for awareness about education. Awareness about facilities and rights is also generated. At present this institution is running a

"Shiksha Karmi Yojna". Under this scheme male or female adult having education upto VIII standard are trained. After training they are deputed in Saharia localities for educating other Saharia. Cultural programmes are also organised by this group.

ASEFA is a voluntary organisation with headquarters in Madras. This institution is working in the area (Kota) since 1979. Since 1987 it is regularly working in the tribal area. It has set up its office at Baran. It is running Adult Education Centres, Sewing and Knitting Centres and Primary schools. Institution deputes two or three persons in a village and these workers organise and conducts all type of development activities regularly, as well as by camps. They arrange medical camps and free medicines are distributed to the Saharia. Camps are also organised to demonstrate modern methods of farming. Saving Group Schemes (Bachat Samuh Yojana) has also been started to cultivate saving habits among Saharia. They also sell food material at a reasonable rate. Some times nourishing food is distributed to children free of charge.

CECOEDECON (Centre for Community Economics and Development Consultants Society) was established in 1981 to provide immediate relief to the victims of devastating floods in Jaipur district. A year later, it was registered as a nonprofit organisation. Its focus shifted from relief and crisis intervention to promotion of sustained community-based economic and ecological development. Over the past years CECOEDECON has undertaken a number of programmes ranging from community-based economic development to peoples empowerment and strengthening of village organisations to land reclamation to nomad development within an ecologically sustainable and bio-regional development context. At the local level, CECOEDECON works with various socially, economically disadvantaged or excluded groups, scheduled castes and tribes and other under-privileged groups like children, women, small and marginal farmers., minorities, landless labourers. At the state level, CECOEDECON offers expertise for undertaking research studies, process documentation and conduct impact evaluation. It also promotes CBOS (Community Based Organizations) by providing training and support in natural resource management, initiating and strengthening public advocacy efforts, promotional support for espousal of people's cause and building coalitions to work on common issues including rights of indigenous people. At the National level the organisation promotes networking and voluntarism with other NGOS and is also associated with Swadhar Foundation (SF), National Centre for Advocacy Studies (NCAS), Voluntary Action Network India (VANI), and National Committee for follow up of World Summit on Social Development (WSSD).

Health Needs Required as told by Community Leaders and NGO Workers

Several discussions were carried out with the community leaders ('Patel'- headman of village and members of 'Panchayat') of various villages and workers of NGOs like Adim Jati Seva Sangh, CECOEDECON (Centre for Community Economics and Development Consultant Society), Lok Jumbish and Sankalp to know what facilities are required at the PHC and Sub-Centres in the villages of Shahabad tehsil. They mentioned that there is a need for X-ray machines, well equiped laboratories (provision for more tests at these laboratories should be made), lady doctor (gynaecologist), specialists (surgeons, orthopaedicians, ENT specialist) at the PHC and Sub-Centres. There is a dire need for supply of medicines. Specialised camps like general check-up, orthoopaedics, paediatric, gynaecological check-up skin camps, eye camps etc should be organised in the interior villages. There is a need of jeeps to carry doctors and the PHC staff in case there is any emergency especially in the interior villages. There is a need for filling up of vacant posts of doctors as well as paramedical staff at Sub-Centres especially in the interior villages.

From Tables 60 & 61 and Figure 15 it becomes clear that the need of opening a hospital or dispensary was more keenly felt in the interior villages of Taleti especially Beelkheradang (55%), Chaurakhadi (50%), Faredua (60%), Pathari (70%) and Sandhri (55%). On the whole 43.5 per cent Saharia of Taleti want opening of hospital or dispensary as compared to 27.5 per cent of Upreti. 20 per cent Saharia of Taleti as well as Upreti each want a doctor to be posted in the villages, 26 per cent Saharia of Taleti and 18 per cent Saharia of Upreti want free and timely supply of medicines. However 4.5 per cent Saharia of Taleti and 14.5 per cent Saharia of Upreti want no other facilities than the existing ones. 6 per cent Saharia of Taleti and 20 per cent Saharia of Upreti did not responded.

Felt Needs of Participants with Regard to Health Services

Focus Group Discussion carried out in six Taleti villages (Faredua, Beelkhera Dang, Chaurakhadi, Matiakhad, Pathari and Sandhri) revealed that all the participants (100%) from these villages felt that a hospital should be opened in the villages especially in the interior villages. There should be regular supply of medicines at PHC and Sub-Centres. All the participants said that there is a dire need for supply of medicines at PHC. Focus Group Discussion conducted in the six villages of Upreti (Shahpur, Gangapura, Mahodara, Dhikwani, Ghesua and Hatri) revealed that all the participants (100%) of all six villages felt that there should be regular supply of medicines at PHC and Sub-Centres because due to their poor economic condition they cannot purchase medicines. They also felt that this was the main reason that

Table 60. Medical Needs Felt by Saharia of Taleti

S. No.	Name of Village	Opening of Hospital or Dispensary	Posting of Doctor in the Village	Free and Timely Supply of Medicines	No Other Facilities	Do not Responded
1	Shahabad	10.0%	—	40.0%	20.0%	30.0%
2	Kusalpura	15.0%	10.0%	50.0%	15.0%	10.0%
3	Matiakhad	30.0%	40.0%	30.0%	—	—
4	Beelkheradang	55.0%	25.0%	15.0%	—	5.0%
5	Chaurakhadi	50.0%	30.0%	20.0%	—	—
6	Faredua	60.0%	25.0%	15.0%	—	—
7	Pathari	70.0%	15.0%	10.0%	—	5.0%
8	Sandhri	55.0%	20.0%	25.0%	—	—
	Total	43.5%	20.0%	26.0%	4.5%	6.0%

Note: Percentage based on 150 Interviews.

According to Jay (1970) there was atleast 100 shamanistic performances and various correlated sacrifices

Table 61. Medical Needs Felt by Saharia of Upreti

S. No.	Name of Village	Opening of Hospital or Dispensary	Posting of Doctor in the Village	Free and Timely Supply of Medicines	No Other Facilities	Do not Responded
1	Shahpur	10.0%	20.0%	30.0%	10.0%	30.0%
2	Gangapura	20.0%	20.0%	25.0%	25.0%	10.0%
3	Samrania	30.0%	25.0%	15.0%	15.0%	15.0%
4	Khandasehrol	40.0%	30.0%	20.0%	—	10.0%
5	Ghesua	40.0%	15.0%	20.0%	—	25.0%
6	Dhikwani	35.0%	30.0%	10.0%	10.0%	15.0%
7	Hatri	30.0%	10.0%	15.0%	30.0%	15.0%
8	Mahodara	15.0%	15.0%	10.0%	25.0%	35.0%
	Total	27.5%	20.0%	18.0%	14.5%	20.0%

Note: Percentage based on 150 Interviews.

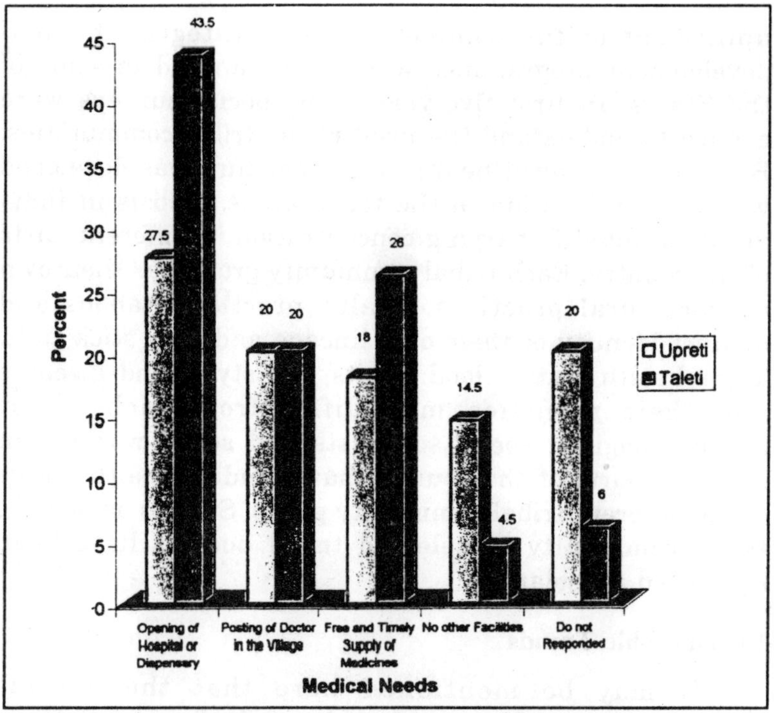

Fig. 15. Medical Needs Felt by Saharias.

treatment of a person was not continued. 50 per cent participants of Upreti reported that health workers do not visit the village regularly and there was no one to approach in case there is any emergency. 75 per cent participants from all these villages felt that proper house to stay was important for good health. All the participants (100%) of the villages felt that adequate amount of food was most important factor maintaining good health. 48 per cent respondents from the villages of Upreti told proper information should be given to users regarding the side effects of contraceptive methods.

A Brief Comparative Account of Status of the Tribal Health

After Independence, the Government of India launched various community development programmes for upliftment of the common masses. Integrated tribal development programmes were also launched in some of the States. In first five-year plan special surveys were carried to understand the need of the tribal communities. Based on that rural health, infrastructure was developed with some relaxation in the tribal areas. Tribals in India are distributed in varying concentration in different parts of the country. Each tribal community group has their own socio-cultural practices, health practices, taboos and distinct identity of their own lineage and they stick to it. Their health status, food habits, life styles and diseases and their remedies may differ from each other. Anthropologists, social scientists and social workers in different parts of the country have studied health status of the different tribal community group. Studies on specific tribal community in selected tribal pockets have been conducted in isolation.

Demographic Trends

It may be mentioned here that the various demographic variables could not be studied among Saharia because of some limitations of the researcher. An attempt has been made to give a broader comparative picture of Saharia with reference to demographic variables. A comparative account of demographic profile of different tribes of India is given in Table 62.

Sex Ratio

It has been observed that there is a decline in sex ratio among the tribal population. It was 972 in 1991 census, which is low as compared to the 1981 census (983). The studies conducted in different tribal population groups of India present a variable picture of sex ratio. Basu et al (1985) found in Jaunsaris of Dehradun district sex ratio of 903 female per thousand males. Sex ratio of 882 females per thousand males was reported by Dutta (1990) in Kora tribal group of Midnapur district of West Bengal. Among

Table 62. Demographic Profile of Different Tribes

Tribes	*Crude Birth Rate*	*Crude Death Rate*	*Sex Ratio*	*Total Fertility Rate*	*GFR*	*IMR*	*Authors*
Saharia (Upreti)	44.3	10.4	903.2	5.04	168.3	82.3	Present Study
Saharia (Taleti)	27.8	12.9	892.8	3.52	113.2	90.0	Present Study
Saharia Total	34.7	11.9	897.1	4.21	137.0	85.9	Present Study
Santal	37.2	10.9	913	5.5	168.0	91.9	Basu (1998)
Lodha	31.2	11.8	902	3.3	133.3	99.17	Basu (1998)
India	32.73	10.1	927	4.30	138.0	77.0	SRS (1992)
Konds (Orissa)	–	–	–	3.6	–	–	Sharma (1978)
Kutia Konds (Orissa)	–	–	–	5.03	–	179.75	Chetlapalli (1991)
Dudh Kharia (Orissa)	38.5	11.8	–	5.39	–	102.4	Kshatriya and Basu (1992)
Desia Kondh (Orissa)	–	–	–	3.89	–	151.28	Chetlapalli et al (1994)
Mirdasa (Orissa)	–	–	–	3.14	–	116.0	Choudhary et al (1994)
Bhils (Madhya Pradesh)	43.5	–	–	–	–	–	Choudhary et al (1976)
Khaiwar (Madhya Pradesh)	–	–	–	4.85	–	–	Sharma and Khan (1990)
Hill Korwas (Madhya Pradesh)	–	9.0	954.4	–	–	–	Panday and Tiwari (1996)
Kota (Tamil Nadu)	–	–	–	3.73	–	–	Gosh (1970)
Sohga (Karnatka)	75.0	–	–	6.16	188.0	–	Murthy (1987)
Jaunswar (Uttar Pradesh)	32.0	11.8	889	3.84	125.0	81.0	Kshatriya et al (1997)
Pnar (Megalaya)	–	–	–	–	–	119.2	Khongsider (1992)

Santals of Midnapur District of West Bengal sex ratio was reported to be 913 and among the Lodha of Midnapur District of West Bengal sex ratio was reported to be 902.9 (Basu, 1998). In Kutia Kondhs, a primitive tribal group of Orissa sex ratio of 920 female per thousand male was observed by Basu et al (1990). There is increase in number of female death and negligence of the female children might have been a contributory factor for decline in sex ratio. The sex ratio among the Saharia of Upreti was found to be 903.2 and 897.1 among the Saharia of Taleti. Low sex ratio among the Saharia can be attributed to their poor utilization of health services, primitive delivery practices and inaccessibility to the health institutions for safe delivery. It has been observed that the sex ratio of Saharia (897.1) is lower than the sex ratio of Rajasthan (910) and sex ratio of India (933).

Fertility and Mortality

High fertility and mortality rates have been witnessed in different tribal communities of India by different workers. Kumar and Mitra (1975) reported high infant mortality and fertility in Tharu of U.P. Basu et al (1985) reported crude birth rate 42.67 per thousand population and infant mortality 79.6 per thousand live births in Jaunsaris of Dehradun district of Uttar Pradesh. Murthy (1990) reported crude birth rate 44.75 among the Soligas of Karnataka. Datta (1990) reported crude birth rate 41.8 among the Koras of West Bengal. Kshatriya (1997) reported a Crude birth rate of 32.0 among the Jaunsaris of Uttar Pradesh. Basu (1998) reported crude birth rate 37.2 among the Santal and 31.2 among the Lodha tribals of Midnapur District of West Bengal. The present study shows Crude birth rate of 44.3 in Saharia (Upreti) and 27.8 in Saharia (Taleti). Infant mortality rate among Saharia has been found to be 85.9, which is higher as compared to Infant mortality rate of Rajasthan (83.0) and India (70.0).

From southern zone, Guru Murthy (1990) reported 25 per cent infant deaths due to diarrhoea and dysentary and 20 per cent deaths due to maternal cause like prematurity, birth injury, multiple birth, low birth weight etc. in Kalyan Dehradun and Belluguppa Blocks of Anandpur District of Andhra Pradesh. In central zone also consisting of Madhya Pradesh, Bihar, Orissa and West Bengal high fertility and high infant mortality has been reported in Jung of Orissa (Ray and Roth, 1991), Kutia Kondh (Basu and Jindal 1990) of Phulbani district, Dudh Kharia of Sundargarh district of Orissa (Basu and Kashatriya, 1992). In Muria (123.25), Marias (85.44), Bhatra (148.56) and Halba (92.78) tribal population groups of Bastar District, high fertility and high infant mortality was found. From Western zone including Gujarat, Rajasthan, Goa, Daman and Diu and Dadra, Nagar and Haveli. Similar incidences of high fertility and high infant mortality has been reported by Sirajuddin (1984), Murthy and Ramesh (1978), Ghosh (1970) and Basu (1967). The infant mortality rate among the Saharia (Upreti) has been found to be 82.9 and Saharia (Upreti) 85.9. Higher Crude birth rate among the Saharia (Taleti) shows that more number of Saharia (Taleti) are adopting family planning methods. Crude birth rate of Saharia (34.7) is the same as that of Rajasthan (34.7).

General fertility rate among the Saharia of Upreti was found to be 168.3 and among Taleti was 113.2. The total fertility rate among the Saharia (Upreti) is 4.85 and 3.35 among the Saharia (Taleti). Age specific fertility rate among the Saharia (Upreti) is 0.97 and among the Saharia (Taleti) is 0.67.The higher fertility rate might be due to lesser number of Saharia (Upreti) were accepting family planning. Total fertility rate among Saharia is 8.2.

Crude death rate among the Saharia (Upreti) is 10.4 and among Saharia (Taleti) is 11.9. Crude death rate among Saharia is 11.9, in Rajasthan is 10.4 and in India is 8.7. The higher crude death rate is due to lack of nutritious food, inadequate knowledge of health care, lack of accessibility to medical and health institution social and cultural practices that are interwoven in their life and unawareness about preventive services.

Concept of Health and Disease Among Different Tribal Population Groups

Although the concept of well being and the notion of the disease varies between different tribal groups, living in different ecological zones with different cultures yet, in a tribal habitat, a person is usually considered to be afflicted with some disease if he/she is incapable of doing the routine work which is usually being expected to be carried out by that individual in the society, i.e., incapacitation from work is the universal index of poor health. Thus the concept of ill health becomes a functional one and not clinical (Bhasin, 1994). This is precisely the reason that among many tribal groups e.g. Kutia Kondha, Muria, Madia, Bhattra, Halba, Jaunsari, Santal, Lodha, Kharia, Bhil, Rathwa, Mina, Jatapu, Saroa, Pando, Khairwar, Oraon, Munda and among many others; symptoms such as pains and aches, weakness, scabies, prolonged cough, mild fever, wounds etc. are not taken seriously as symptoms of disease.

Like other tribes in India the Saharia also believe in supernatural powers and they have strong feeling that different deities and spirits are responible for causation of different diseases. The spirits of dead ancestors are also believed to play an important role in ensuring good health, prosperity and protection to the community. Among Saharia, there is a strong belief that persons emanating evil spirits, having mystical powers like evil eye and evil touch can also affect or deteriorate the health of a person. There are traditional healers ('Janteyar's) in the community who identify the cause of illness and treatment is based on it. The traditional healer ('Janteyar') develops faith and assurance for good health in the patient and treats him. They do not associate disease and epidemics with their insanitary condition of living but associated them to the wrath of the Gods and Goddesses. There is no concept of nutritious diet having good effect on health among the Saharia. Physical isolation, ecological legacy, traditional culture and beliefs have made the Saharia self-reliant and they have developed protective mechanism for their survival. Majority of Saharia especially in the interior villages is ignorant about the modern amenities and still follow the rituals of pleasing supernatural powers.

In some of the tribal communities like Santhals of Mahurganj District milk is not consumed and is thrown away as it is not good for human consumption (Ray and Roth, 1991). Tribals spend their limited resources on pacifying spirits of deities and Gods and Goddesses, whom they consider are responsible for diseases and natural calamities. The treatment procedure in tribal includes preventive and curative aspects. In preventive aspect, they perform rituals, use charms and amulets and animal sacrifice whereas in curative aspect they worship deities/spirits and take the help of sacred specialists, priests and medicine men. In some tribal communities religious performance is performed for treatment of plague, chicken pox etc. which are supposed to be caused by supernatural powers. Tribal community is very homogenous group and treatment of the diseased person is the community responsibility. Illness is not considered as individual affair but from the selection of the health provider to treatment of the disease is decided by the community.

A tribal in general, hardly makes a distinction in the magnitude of fever. However, within the limits of their own respective world view, most of the tribal societies have definite means for identifying and classifying various kinds of ailments and diseases. It may be worth while to state that at least one component of health is universally seen among the tribal societies, and that is, committing or omitting certain acts. In other words a breach of trust is thought to bring upon some kind of affliction on the individual or a family as a whole. Measles, tuberculosis, diarrhea, cholera are some such diseases where individual's action may cause some concern to the family, clan or the village. Interestingly, the causation of such diseases is independent of the sanitary condition of the community/individuals. The fate of the individual and the community depends on their relationship with unseen forces that intervene in human affairs. If human beings offend them, the mystical powers punish by causing sickness, death or other natural calamities. The tribal people believe in the presence of benevolent and malevolent spirits, the former playing a protective role, while the latter are considered to be responsible for causing disease and epidemics. It is a common observation across the tribal culture that the ancestral spirits play important role in the prosperity and protection of the family. And they have to be properly honoured for otherwise they will bring wrath on the family members. These spirits are believed to bring a state of physical, mental and social well being to the members of the family. There are many spirits who are feared because of their power to afflict people by bringing a number of diseases. The role of spirits, ghosts and dieties in the tribal life in the causation and treatment of disease is so important that the local tribal people have to seek the help of traditional diviners, medicine man, sorcerers (Sirha, Gunia, Bhua, Janteyar, Bhopa, Ojha, Pujari etc.) for appeasing, controlling or driving away the disease causing agents. The frequency of worshipping or getting in touch with malevolent spirits is more common because of their immediate effect on day to day life.

Dave (1960) conducted study on Garasia and reported that they have faith in indigenous herbs found in the forest around their habitat. In case of ailment, an elderly Garasia, who is known to have knowledge of various herbs, is consulted.

Opler (1963) listed seven major causes of various diseases, viz., malfunctioning of 'doshas', faulty diet, lack of harmony with supernatural world, activities of ghosts, imbalances of forces which control health, in appropriate behaviour in physical, social and economic matters.

Hasan (1967) distinguishes between two types of social and cultural factors that affect the health of any community: (a) factors that directly affect the health of the community because certain customs, practices, beliefs, values and religious taboos etc., create an environment that helps in the spread or control of certain diseases and (b) factors that indirectly affect the health of the community as they are related to the problem of medical care to the sick and the invalid. A physician who does not try to understand that cultural and intellectual level of the people and does not develop respect for cultural differences, proves to be a less successful practitioner in the village. If scientific medicine is to be carried to rural areas it is necessary for medical practitioners and public health workers to acquire adequate knowledge of the general concepts of culture and social organisation of that community.

According to Jay (1970) there was atleast hundred shamanistic performances and various correlated sacrifices per year in the tribal village of central India taking into account only family related rituals.

Banerjee (1988) has pointed out that Gonds have their own traditional medicines and cures which are effective enough in ordinary diseases, Gonds have more faith in magic cures done by 'gunia' and 'panda'. They believe that sickness in most cases is caused by evil spirits or by the wrath of gods, who are not worshipped or are neglected. The villagers are still under the influence of 'ojha'. For the treatment of boils etc., they practice a gruesome method of pricking them with burning needle.

Singh (1994) examined the interplay of several factors influencing health of tribal communities namely (a) physical environment/elemental, forces of nature like earth, sun, wind, rain, forest, (b) Socio-economic state (c) Nutritional availability and dietary habits (d) Psycho-social culture (e) Health culture and Health related behaviour (f) Mortality and Morbidity pattern (g) Genetic diseases and disorders, (h) The therapeutic system in vogue and (i) Health delivery systems. Singh pointed out that issues concerning tribal health, nutrition and gentico–environmental aspects are grossly under researched.

It has been realised by many workers that diseases are not only due to physical, chemical or biological processes but also due to a number of socially and culturally determined factors (Swain, 1994; Basu, 1990).

Few studies have been carried out to throw light on indigenous medical beliefs and practices and western medicine; treatment with religious rites; health services; impact of development programmes on tribal health; role of allopathic doctors in tribal areas; utilization of modern health practices and delivery system; pattern of health seeking behaviour (Singh et al, 1971; Jaggi, 1973; Banerjee, 1982; Rizvi, 1982; Gupta, 1986; Mann and Mann, 1986; Khan, 1986; Choudhary et al 1986; Banerjee, 1988; Bardhan et al, 1989; Ali, 1992; Basu et al, 1994; Sinha, 1994) whereas some of the studies (Roy Burman, 1990; Mahapatra, 1994; Bhattacharya, 1995; Ghoshmaulik, 1995; Kapoor, 1998;1999) have recently touched upon the other dimensions of health which have been discussed.

Mahapatra (1994) studied the nature of linkage between traditional medicine and modern system of medicine and also the mechanism of exorcisms, spirit healing and medicine preparation.

According to Bhattacharya (1995) some indigenous medicinal plants have been reported for the treatment of veneral diseases i.e., Gonorrhoea, Syphilis among the tribals (Bharia and Gond). It has been observed that these tribals use indigenous plants for the treatment of veneral diseases.

Ghoshmaulik etal (1995) described the health situation of the Juangs, a primitive tribal group of Orissa. The disease prevalence among the Juangs suggest that these people are mostly victims of environmental infections. The study reveals that the Juang people are more dependent on the medicine man than the government primary health centre. The village indigenous medicine man or Shaman receives respect and faith, as the villagers think that he treats them well. The people visit the village medicine man more frequently because the primary health centres are located far away from most of the villages. It was found that the local health centres are not fully equipped with required medicines for the general treatment of the people of the villages.

Kapoor (1998) examined the concept of health, treatment and faith in a scheduled tribe of Himalayas and reported that the tribe has got more faith in traditional practices rather than modern. Kapoor (1999) reported the concept of disease and its healing aspect in various caste groups of Himalayas.

It has been observed from various studies that health and treatment reflects the social solidarity of a community. In a tribal community, for example, illness and the consequent treatment is not always an individual or a familial affair, but the decision about the nature of treatment is taken at the community level. In the rural areas, in case of some specific diseases, not only the diseased person or his/her family, but the total villages community is affected. All the other families in the village are expected to observe certain taboos or norms and food habits. The non-observance of such practices often calls for action by the village council. One cannot deny the impact of this psychological support in the context of treatment and cure. In some societies, illness may be an individual affair, and the patient may be looked after by the administration (state), while in tribal societies it is a familial or sometimes community affair where necessary action and care is taken by the community as a whole.

It has been observed that the dependence on and confidence in traditional medicine-man or magicians and shamans are again often responsible for the non-acceptance of modern medicine. The traditional approach establishes faith and assurance in the patients, while modern medicine lacks this. As the shamans or what are called traditional medicine-man and magicians share the common cultural beliefs and practices of the patients, naturally they have more faith in them (Buchs, 1964; Dube, 1970).

A few studies (Khan, 1986; Basu, 1994; Mahapatra, 1994; Sinha, 1994) have been conducted to study the interaction between traditional and modern system of medicine among the tribals.

There are some studies (Khan, 1986; Mahapatra, 1994; Sinha, 1994) which have shown that when both the facilities namely, traditional and modern, are available, the tribals often accept the modern system.

It is been noted by many that tribals living in remote areas have a better health status and more balanced food than those living in less remote areas (Chaudhuri, 1986; Chaudhuri etal, 1989). Roy Burman (1990) has noted that physical well-being of the tribals is inextricably linked up with the forest ecology and rapid disappearance of forest, commercialization of minor forest produce, replacement of food crops by cash crops and tendency towards monoculture forestry and lastly privatization of communal lands may affect the health and nutrition and physical well-being of the tribal population.

The studies conducted on different tribal groups living in different ecological conditions shows that physical isolation, adverse climatic conditions, ecological legacy and traditional culture and beliefs have made them self-reliant and they have developed protective mechanism for their survival. However, the notion that tribals don't accept the modern system is not true. The tribals do accept the modern system if the facilities are made available to them.

Maternal and Child Health Status

Maternal and child care is an important aspect of health seeking behaviour. Maternal malnutrition leads to premature deliveries, low birth weight babies and pregnancy wastages. Data on this aspect is fragmentary on the tribal population groups of India. Nevertheless, the trends that are coming up are alarming. In one of the study on Bhotia, Kapoor and Prasad (1982) have observed that in the event of delayed conception of married women, the deities like Nanda Devi, Nabling and Kalika are propitiated by taking on earthen lamp in a plate for the whole night. This sort of appeasement for the procurement of child is known as 'Jagran' or 'Dardeo'. There is a popular belief among the Bhotias that a child born during the 'Amawasya' (dark phase of the moon) causes hardships to himself or his parents.

Prasad and Kapoor (1982) discussed the obstetrical practice among the Bhotia tribals. The Bhotias use very simple and unhygenic method to expell the placenta, locally known as 'Phali', from the mother's womb after child birth, if it is not expelled in natural course. The Bhotia women are also given a gentle abdominal massage and tea brewed from the 'Bodpa' leaves, found in Tibet, to effect easy delivery of the child.

Banerjee (1988) has mentioned that the Gonds, after the birth of the child, cut the umbilical cord with an unsterilized sickle and this often leads to tetanus. There has been no orientation programmes for the Dais.

In Bastar District (Basu and Kshatriya, 1989) it was observed that maternal and child care was largely neglected. Expectant mothers to a large extent are not inculcated against tetanus. From the inception of pregnancy to its termination, no specific nutritious diet is consumed by women. The consumption of iron, calcium and vitamins during pregnancy is poor. More than 90 per cent of deliveries are conducted at home attended by elderly ladies of the household. No specific precautions are observed at the time of conducting deliveries which result in an increased susceptibility to various infections. Services of paramedical staff are secured only in difficult labour cases. It has been observed that a specially prepared concoction from various kinds of herbs known as "Kasa Pani" is provided to mother on the third day after delivery, "Kasa Pani" is a herbal medicine which according to them gives soothing effect, helps in recovering the depleting energy and also acts as anti-infective. Maternal mortality directly related to pregnancy and childbirth has been found to be appreciably high among the tribal populations of Bastar district. In addition, a lot of females suffer from ill health due to pregnancy and childbirth in the absence of well defined concept of health consciousness.

Devarapalli (1992) studied maternal care and obstetric practice among the Konda Dora tribes of Andhra Pradesh. It was found that Konda Dora women continued to work till the time of delivery. Dietary restrictions were also imposed on her during pregnancy. Beans were avoided as they were believed to cause epileptic fits. Deliveries were conducted at home. Umbilical cord was cut with the help of sterilized knife. After delivery the women remains completely at home for a minimum of one month.

Prema and Thomas (1992) studied the nutrition and health problems faced in normal and and physiological conditions like pregnancy and lactation among the kannikar tribal women of Kerala. Pulses, milk and milk products and other animal foods which were the main sources of protien were lacking in their diets.

Pandey and Tiwari (1993) carried out their studies on Hill Korwas of Madhya Pradesh. Of the fourteen women, who were pregnant at the time of interview, only one had consulted ANM during her pregnancy. Only five seemed to know about tetanus toxoid injections and intake of iron tablets during pregnancy. Most of the deliveries were conducted by untrained dais or traditional birth attendants (7·2 per cent) followed by elderly ladies (23.9 per cent) and less than one per cent were conducted by trained dais and the ANM. The cord was cut with a bamboo piece for a male child and with a knife for female child. The colostrum was not discarded. Goat milk was given as a substitute for mother's milk.

Pandey et al (1996) studied the maternal and child practices among the Khairwar women of Madhya Pradesh. The findings showed that pregnant women did not receive any special antenatal care or diet. There was no restriction on her daily work, including fetching of water and cutting of wood. The delivery was usually conducted by a elderly woman and the ANM or trained Dai was not called and in case of complication the pregnant women was usually left at the mercy of god. The umbilical cord was usually cut with the help of sickle or old blade.

The chief causes of maternal mortality were found to be unhygienic and primitive practices for parturition. For example, it was observed that among Kutia Kondhs (Basu et al., 1990), the delivery was conducted by the mother herself in a half squatting position holding a rope tied down from the roof of the hut. This helped her in applying pressure to deliver the child. In complicated labour, it might lead to maternal as well as child mortality. Similar crude birth practices were found to exist in other tribal groups like Kharias, Gonds, Santals etc.

As far as child care is concerned, both rural and, tribal illiterate mothers are observed to breast feed their babies. But most of them adopt harmful practices of discarding of colostrum, delaying the initiation of breast feeding and delaying the introduction of supplementary feeds. Vaccination and immunization of infants and children have been inadequate among tribal groups. Since the personal hygiene is very poor, the under five children are the worst sufferers and most vulnerable to infections. In addition, extremes of magico-religious beliefs and taboos tend to aggravate the problems.

Among the other morbid conditions, yaws, leprosy, tuberculosis and malaria have been found to be quite prevalent in tribal areas. Yaws is caused by Treponema pertenue and cases have been reported from (Bastar) Madhya Pradesh, Tripura and Maharastra (NICD, 1993). Incidence of tuberculosis has been found to be increasing in tribal areas as the tribals are coming more in contact with other non-tribals. Poor nutritional status is also one of the contributory factors.

Incidence of leprosy has been reported in scheduled tribes of Bihar, Madhya Pradesh, Orissa, Uttar Pradesh, Tripura, Incidence of malaria is wide spread in almost all the tribal areas. Tribals who constitute 8.1 per cent of the total population contribute 25 per cent of the total malaria cases in the country. About 50 per cent of the P.falciparum cases reported in India come from tribal population. Malaria is major health problem in all the tribal areas.

Very little information is available about the prevalence of non-cummunicable diseases in tribal population. Studies carried out by NIN (Rao et al, 1986) among the tribals of Sarguja, Jhabua and Bastar districts of Madhya Pradesh showed high incidence (53.9 per cent) of Goitre (enlarged thyroid gland) in tribals of Sarguja district. Eye conditions like cataract, corneal opacity etc. were found to be in high frequencies (20.3 per cent) in Bastar district.

Nutritional anaemia is quite common in women and children in the Saharia area as in other tribal states. The diet taken by Saharia does not contain adequate amount of Vit. A, calcium, Vit. C, and animal protein. During the period, of pregnancy the Saharia woman is required to observe a number of taboos. For example, She must not look at the sun or the moon during eclipses failing to do it would disfigure the baby in the womb. If the delivey is delayed, offerings are made to family goddess known as 'Kodia Devi'. The old women of the family or neighbourhood help in the delivery. A 'Bhangan' (sweepress) is called to cut the unbilical cord ('Nal Katna'). The Saharia consider that it is only the work of the 'Bhangan' (sweepress). In the interior villages there is no sweepress and the sweepress is called from other village which requires two to three days and this has an adverse effect on the health of the mother and the newly born baby. The 'Bhagan' cuts the umbilical cord with blade and knife. The placenta is burried in the cattle shed. Basu etal (1989, 1993) reported high incidence of angular stomatitis and mottling of teeth in Gond children of Bastar District. Prema and Thomas (1992) in their study of Kannikar tribal women of Trivandrum District, Kerela found very high incidence of anaemia (90%) Vit. A deficiency and (30%) and Niacin deficiency (10%) on clinical examination of normal and physiological conditions like pregnancy and lactation. There is no correct statistical information available on incidence of maternal mortality in tribal population groups but some authors have reported very high maternal mortality in tribal population groups. Tribal women do hard manual work upto the last trimester of pregnancy. There is no concept of nutritive diet during pregnancy and alcohol intake is quite common during pregnancy. More than 90 per cent of deliveries are conducted at home. These practices are responsible for poor maternal and child health status. Immunisation against tetanus and vaccination to children was found to be highly inadequate in tribals. In study of Mundas and Oraons of Ranchi District, Singh et al. found 44 per cent children had severe malnutrition. Girls have higher incidence of severe malnutrition (47%) as compared to the boys (42%). The daily consumption of milk, fish meat was less than 1 per cent. The data on incidence of reproductive wastage and still birth has been reported in selected tribal population of certain states. In Hill Kolkan tribals (Murthy and Ramesh) of Andhra Pradesh, incidence of stillbirth was 3 per cent and abortions were 1.4 per cent. In Murias of Madhya Pradesh stillbirths were 0.05 per cent and abortions were 0.05 per cent (Saha, 1982). In Zemi tribals of Nagaland Chaudhary et al. (1971) reported stillbirth rate of 4.35 per cent and abortions 7.5 per cent. In Kondhs of Orissa, Sharma (1979) reported the incidence of stillbirths was 2.17 per cent. Primitive delivery practices and inaccessibility to the health institutions for safe delivery may be one of the contributory factors. In a study conducted by RMRC Bhubaneshwar in tribal dominated Districts of Sundargrah, Koraput, Mayurbhanj and Keonjhar 40 per cent illness among children was found to be due to dysentry, upper respiratory tract infection, scabies, malaria and anaemia. Child mortality at age of two was highest in Koraput and Keonjhar and lowest in Mayurbhanj. The mortality rate rose from age two to five years. It was 27 per cent in Sundergarh, 25 per cent in Mayurbhunj, 16 per cent in Koraput and 11 per cent in Keonjhar.

Thus it is very evident that maternal and child health scenario among the Saharia and other tribes of India living in different ecological conditions is very grim.

The above mentioned studies carried out on different population groups of India show that health status of the tribal population in India is very poor. Tribals in India are distributed in varying concentration in different parts of the country. Each tribal community group has their own socio-cultural practices, health practices, taboos and distinct identity of their own lineage and they stick to it. Their health status, food habits, life styles and diseases and their remedies may differ from each other. Every tribe is living in a different ecological zone, has a different culture and has different health care practices. But the health status of the tribals living in different ecology with different cultures present a dismal picture. The present study shows that health status of Saharia is poor. Numerous scholars have tried to establish poor health status of tribals with the help of morbidity, mortality and health statistics. The widespread poverty, illiteracy, malnutrition, absence of safe drinking water and sanitary and living conditions, poor material and child health services, ineffective coverage of national health and nutritional services. The present study on Saharia have also traced out factors like widespread poverty, illiteracy, malnutrition, absence of safe drinking water and sanitary and living conditions, traditional health care practices, superstitions, totems and taboos, poor maternal and child health services, lack of health infrastructure, inaccessibility of modern medicine to tribals, ineffective coverage of national health and nutritional services which are responsible for dismal health conditions prevailing among the Saharia. The present study shows that there exists a relationship between ecology, culture and health. It can be said from the present study that poor health status of Saharia reflected in their health care practices is the outcome of their ecological, social and living conditions, illiteracy, ignorance, lack of transport facilities, under availability of potable water etc. and in appropriate medical and health opportunities.

6

Summary, Conclusion and Suggestions

Introduction

Medical anthropology is a bio-cultural discipline concerned with both biological and socio-cultural aspects of human behaviour, particularly with the ways in which the two interact and have interacted throughout human history to influence the conceptions of health and treatments and therapies of disease (Foster and Anderson, 1978:2-3).

The tribal groups of India form about 8.1 per cent of the total population of India. There are 75 Primitive Tribes in India. Most of the researches have been carried out on different Primitive Tribes of India to collect ethnography (Roy,'25, '35; Majumdar, '33, '37, '61; Griegson, '38; Elwin, '39, '50; Khan,'42; Haimendorf, '43, '45, '48; Aiyappan, '48; Dube, '51; Nag, '58; Srivastava, '58; Jay, '59; Bhowmick, '63; Naik, '63; Vidyarthi, '63, '81; Dasgupta,' 64; Sinha,' 68; Prasad, '81; '93; Sachidananda, '81; Adhikary,' 92). Mostly these studies have put emphasis in the field of religion, kinship, family, marriage and social change in Primitive Tribes of India. These studies are isolates and fragmentary in nature.

In India, life styles, dietary pattern, social behaviour as well as health behaviour are prescribed by the deep rooted traditions. These are more pronounced in the tribal areas where outside influences, education and urbanisation are minimal (Chattejree, 1993). The health seeking behaviour of the Primitive Tribal Groups (PTGs) is supposed to be highly associated with their beliefs, customs and practices, since, majority of the population in these societies are illiterate, poor and lacking modern health facilities. Health and disease are a continuous process and are interlinked with the health seeking behaviour of the community. Zola (1964), Mechanic (1978), Rosentock (1960), Pandey and Tiwari (1993) each have identified a number of variables to the health seeking behaviour of the individuals, some of the common ones are lifestyle of the person, the seriousness of the condition, the tolerance threshold of the patient, the frequency of recurrence, the financial demand, availability of treatment resources and above all the cultural pattern of the community. The conceptualisation and measurement of health and quality of life are gaining increasing attention in the health services. One of the major issues in the health status measurement is the health seeking behaviour of a community which governs the morbidity and mortality pattern. Since the concept of health occupies different meaning in different social systems, the health seeking behavior of a community cannot be studied in isolation from the social network of a community as it is deeply interwoven into every event of social, economical and biological aspects of a population. Each of these aspects has a deep influence on health and to raise health status and quality of life of people sustained efforts are required to bring about transformation of the society as a whole in graded amplitude. Some attempts have been made to study the health seeking behaviour of different tribal communities of India (Pandey and Tiwari, 1993; Basu, 1994; Singh, 1994; Swain, 1994).

Health and culture are very much connected with ecology. The traditional health care system and treatment are based on deep observation and understanding of nature and environment. The hypothesis of the present study was to ascertain whether there is an interrelationship between socio-cultural practices, environmental conditions and health aspects exists among Saharia. The specific aims and objectives of the present study are:-

1. To study the ecological, demographic, cultural and economic aspects affecting the health of the tribe.
2. To understand the concept of health, illness and nature of treatments and to identify the pattern of health seeking behaviour.
3. To study the impact of various health programs on the tribe and
4. To ascertain their preference for type of system of medicine and to identify factors associated with non-utilization of health services provided by government agencies.

As there is no study on any primitive tribal group emphasizing on ecological, cultural and health aspect, this health oriented action research study will help in formulating need based health care strategy for the Saharia in particular and other tribes of Rajasthan in general.

There is paucity of comprehensive health related studies among the tribal population. Most of the studies are isolates and fragmentary in nature. There is urgent need for initiating area specific, group specific, health need action research in consonance with the felt needs of the tribal communities. In India, studies of health behaviour, treatment process and health culture of the tribal people are of recent origin. It is quite well-known that in India there exist not one but a plurality of health care systems and norms, beliefs and practices regarding health and treatment process. The folk medicine is on one side of the spectrum while on the other side there exists modern scientific medicine system. Thus in contemporary India not only plurality of health cultures exist but there also exist different health care systems as well. The health culture of tribals has attracted the attention of anthropologists. However, the studies conducted by anthropologists on tribal health emphasize rather over emphasize the cultural dimensions of health. Moreover these studies suffer from the lacuna of treating the cultural factors in isolation. Cultural factors reveal only partial truth. They can explain to some extent the health culture of tribals. But this is not all because many factors interact and impinge upon health. Health is as much a cultural phenomena as it is a reflection of the political, economic, environmental and social forces prevailing in a tribal community. Health is very intimately connected with ecology and culture. Therefore, this study will help in understanding the health problems and health care system of community using holistic and interactive approach.

Area and People

The State of Rajasthan is situated in the North Western part of the Indian Union (23°30‘ and 30° 11‘North Latitude and 69° 29‘and 78° 17‘ East Longitude). It's total area is 3,42,239 sq. kms. The forest area of Rajasthan is 9 per cent of its total area. The population of Rajasthan is 4.40 crores which is 5.20 per cent of the nation's population. The density of population in the state is 129 person per sq. kms. compared to 273 for India. The sex ratio in Rajasthan is 910 as compared to 927 of India. According to the 1991 census, the tribal population in Rajasthan is 54,74,881, out of which 28,37,014 are males and 26,37,867 are females. It is nearly 8.10 per cent of the total tribal population of the country and 12.44 per cent of the total population of Rajasthan Of the total tribal population in Rajasthan, 9.35 per cent live in rural areas and the rest 4.65 per cent in cities.

As regards the geographical distribution and an understanding of the demographic situation of the tribals, the Rajasthan State may be classified into three regions (southern region, south-eastern region and western region) on the basis of concentration of tribal population. The southern region has the maximum tribal concentration in the state, i.e. about 59.9 per cent of the total tribal population of the state. The major Districts of this region are-Udaipur, Dungarpur, Banswara, Chittorgarh, Sirohi, Pali, Bhilwara and Jhalawar with a tribal population of about 36.79 per cent (10,63,071), 65.84 per cent (5,75,805), 73.47 per cent (8,49,059), 20.28 for cent (3,00,971), 23.39 per cent (1,53,005), 5.39 per cent (80,265), 9.26 per cent (1,43,748) and 11.89 per cent (1,13,834) of the total population of these districts respectively. The main tribes of this region include Bhil, Bhil-Meena, Damor, Garasia and Kathodi. The Districts of Dungarpur and Banswara and the Pratapgarh tehsils of Chittorgarh District of this region have been declared as 'scheduled area'. The south-eastern region is a sparsely populated area with a tribal population of about 32.12 per cent of the total tribal population of the state. It includes the Districts of Alwar, Bharatpur, Sawai Madhopur, Kota and Baran, Bundi, Jaipur and Tonk with a tribal population of 8.05 per cent (1,85,048), 2.31 per cent (38,212), 22.58 per cent (cent (4,43,469), 14.19 per cent (2,88,367), 20.25 per cent (1,56,001), 11.25 per cent (5,31,652) and 11.89 per cent (1,15,948) of the total population of these Districts respectively. The important tribes of this region are Meena, Saharia, Kathodi, Patelia, Naika and Koli. The western region is an arid region and is inhabited by the minimum of population i.e. only 7.98 per cent of the total tribal population is living in this area. It comprises the Districts of Barmer, Jalor, Jodhpur and Sikar with a tribal population of about 5.86 per cent (84,232), 8.43 per cent (96,324), 2.82 per cent (60,819) and 2.64 per cent (48,667) of the total population of these Districts respectively. The important tribes of this region are Dhanka, Kokna and Naika. The tribal population in Rajasthan is thus concentrated in belts running from Sirohi through the Udaipur, Dungarpur, Chittorgarh and Banswara Districts to the Bundi, Kota and Baran, Sawai Madhopur, Tonk and Jaipur Districts. The borders of Banswara, Dungarpur, Udaipur and Sirohi Districts meet the Sabarkantha, Banaskantha and Panchmahal Districts of Gujarat State, while the Saharia populated Shahabad and Kishanganj tehsils of Kota and Baran Districts share a common boundary with Madhya Pradesh. The present study was conducted on Saharia of Baran District of Rajasthan. The Saharia are the only Primitive Tribe of Rajasthan and inhabit in the Shahabad and Kishanganj tehsils of Baran District. The total area of Baran District is 2898 sq. km. Shahabad has a total area of 1469 sq. km whereas Kishanganj has an area of 1429 sq. km. The number of villages in Shahabad tehsil are 236, out of which only 146 are inhabited villages and rest of the villages are uninhabited. The total population in Shahabad tehsil is 83,028.The population of Saharia in Shahabad tehsil is 28,632 which forms 34.48 per cent of the total population. The population of Saharia in Kishanganj tehsil is 19,190 which forms 17.71 per cent of the total population. The number of villages in Kishanganj tehsil are 203, out of which only 171 are inhabited villages. The District Baran contributed 1.84 per cent to the total population of the State. Density of population of Rajasthan works out to be 129 sq. km whereas it is only 116 persons per sq. km in case of Baran District. To cater to the needs of Saharia, special programmes were started in 1977-78 for the development of Saharia (Primitive Tribe) residing in Kishanganj and Shahabad Panchayat Samitis of Baran

District (formerly Kota District). According to Saharia Development Project Area (1991), the total population of Saharia in Baran District is 47,822.

The Saharia of Rajasthan are concentrated in two tehsils of Shahbad and Kishanganj which also coincide with their respective Panchayat Samitis. The Shahabad and Kishanganj tehsil can be divided into two distinct divisions (1) Upreti or Upland and (2) Taleti or Lowland. The Upreti constitutes Kishanganj teshil and the western part of Shahabad tehsil. The remaining part of Shahabad tehsil comes under the classification of Taleti.

Methodology

The extensive and intensive field work was carried out in 16 Saharia villages (Shahabad, Kusalpura, Shahpur, Gangapura, Khandasehrol, Ghesua, Hatri, Dhikwani, Mahodora, Samrania, Devri, Faredua, Chaura Khadi, Beelkhera Dang, Pathari and Sandhri) for 8 months from Sep1999 to April 2000. Data was collected from 300 households (150 households from Taleti and 150 households from Upreti). In order to find out the relationship between ecology, culture and health among the Saharia both primary and secondary data was collected. Both qualitative and quantitative approaches were used for the collection of primary data. Qualitative techniques used included Observation (participant and non-participant), interviews, case study method and Focus Group Discussion (FGD).

Quantitative data was collected with the help of a Household schedule and a Village Profile schedule.

Secondary Data was collected from office of District Collector (Baran), Adim Jati Sewak Sangh (Shahabad), 'Saharia Vikas Pariyojna' (Shahabad), Lok Jumbish (Shahabad), Census of India (1991), PHC (Shahabad), Information Centre (Kota).

Socio-Cultural Life of Saharia

Saharia do not have disperse pattern of living as found among the Bhils of Southern Rajasthan. They do not have even separate and scattered hamlets. However, the hamlets are slightly far off from the main villages. Each village has a cluster of Saharia families in a separate hamlet known as 'Saharana'. In the Saharana no other caste group live. In the centre of each Saharana they construct umbrella type shelter house which is called 'Panchayat Bungalow' where all Saharia sit together to discuss, gossip and settle their disputes.

Saharia is an endogamous group divided into exogamous clans and these are named after some ancestor or a place to which they originally belong

Saharia believe in totem and taboos and observe them during rituals and ceremonies of marriage. Marital alliance between the mates of same clan is prohibited by virtue of their affiliation to a common totem. Widow marriage or remarriage called 'Nata' is permissible among the Saharia. The family of Saharia is a nuclear family (98%). In the present study it has been found that the literacy rate in Saharia (Upreti) is 12.1 per cent as compared to 8.3 per cent in Saharia (Taleti).

The religion practiced by Saharia is Hinduism. They worship Hindu gods and godesses like balaji, bheruji, ambamata, bijamata. They have faith in supernatural powers. They also believe in ghosts and spirits.

The fairs, festivals and rituals observed by them come close to Hindu fold. They celebrate Teja Dashmi, Navratra, Makar Sanskranti, Deepawali and Holi with great joy and enthusiasm. Fairs have an important place in the life of Saharia. All Saharia assemble in Sitabari fair. Taking bath in water reservoirs at Sitabari is considered a religious ritual. Attending Sitabari fair is considered a duty. They believe that Mahrishi Valmiki resided over here and Sita also lives here in his Ashram.

HEALTH STATUS AND PRACTICES

Overview of Baran District

District Baran is situated at a distance of 72 km by road and 67 km by train from the District Kota. The Saharia are the only Primitive tribe of Rajasthan and inhabit in the Shahabad and Kishanganj tehsils of Baran District of Rajasthan. The District Baran has a dry climate. The principal river of the District is Chambal which has a perennial flow. The District has two hospitals, four Community Health Centres (CHCs), eleven Primary Health Centres and fifty two Government dispensaries. There are 912 in-patient beds in two hospitals of the District. Shahabad tehsil is the centre place of the District. The total population of Saharia in eight villages of Upreti is 3583 (1882 females and 1701 males) and 5675 in eight villages of Taleti (2996 males and 2679 females). The villages under study lacked basic essential facilities like banks, post offices, cinema houses. There is no chemist shop in the interior villages.

POPULATION PROFILE

Soci-Economic and Environmental Characteristics

All the Saharia houses (100%) were Kaccha houses. All houses had only one room. 59.7 per cent households had less than five Bighas of land. The main source of income of Saharia include agriculture (43%), forest produce (16%) and labour at construction sites (41%). About 71 per cent households obtain their drinking water from the hand-pumps, 15.7 per cent from rivers, 13.3 per cent from the wells. The water tanks were found empty. 84 per cent households had no electricity. None of the households had toilet facility. All the households used earthen smoke ovens (chulhas). Wood was used for fuel in all the houses. Sanitation among the Saharia is very poor. It has been observed that the Saharia go to the field for defecation. The existing habits of defecation may be responsible for ill health of Saharia in those areas. Personal hygiene is also quite poor among the Saharia. This is due to poor awareness about the importance of personal hygiene. 42 per cent Saharia had income less than Rs 500/-, 32 per cent have income in the range of Rs 500/- Rs 999/- and only 5 per cent have income in the range of Rs 2000/-Rs 2999/-only.

Demographic Trends

The population of Saharia (Upreti) in the present study has been found to be 767(403 males and 364 females) whereas in Saharia (Taleti) the population composed of 1077 (569 males and 509 females). It has been observed that Saharia is a growing population as 38.45 per cent of the population has been found in 0-14 years age group. Saharia of Taleti show higher i.e. 40.85 per cent than the Upreti (35.07 per cent) in 0-14 years age group. It appears that Saharia of Taleti are growing at faster rate than Upreti. The total dependency ratio (TDR) among Saharia has been found to be 67.33 and Young age dependency ratio (YADR) and Old age dependency ratio (OADR) has been found to be 64.34 and 2.99 respectively. It has also been observed that Saharia of Upreti show lower dependency ratio (59.46) and young age dependency ratio (55.93) than Saharia of Taleti region. OADR show slightly higher in Upreti than Taleti region among Saharia. The index of ageing among Saharia is found to be 4.65 where it is higher in Saharia of Upreti than Taleti. It has been found that the age-cohort wise sex ratio of Saharia (Upreti) favours the males in the age groups from 0-14 years where as it favours the females from the age groups 15- 29 years. Then again it favours the males from 30 years to 49 years and favours females from 50 years onwards upto 70 years. The overall sex ratio among Saharia (Upreti) has been found to be 903.2. It has been found that the age-cohort wise sex ratio of Saharia (Taleti) favours the males distinctively in the age groups from 0-14 years whereas it favours the females in the age groups 45- 49, 50-54, 60-64 and 70 years respectively. The overall sex ratio among Saharia (Taleti) has been found to be 892.7. The sex ratio among Saharia has been found to be 892.7. CBR of Saharia has been found to be 34.7/1000 which is equal to Rajasthan and is more than national average. Saharia of Upreti show much higher CBR (44.3/1000) than the Taleti. Saharia present high CDR of 11.9 than the Rajasthan (10.4) and national average (8.7). Saharia of Taleti show much higher CDR (12.9) than the Upreti (10.4). General Fertility Rate (GFR) has been found to be 137.0 in Saharia whereas Saharia of Upreti show higher General Fertility Rate (168.3) than Saharia of Taleti. It has been observed that in both the regions of Saharia i.e. Upreti and Taleti, the age specific fertility rates have been found to be increased from the age-groups of 15-19 years to 25-29 years. There after age specific fertility rates decreased gradually showing a little and uneven trends at later years in both the regions of Saharia. The figure shows that the number of children to be expected for 1000 married women in their reproductive life has been 841.42. Total Fertility Rate (TFR) is estimated at 4.21 births per woman among Saharia which is higher than national average. It has been observed that the total fertility rate (TFR) is higher in Saharia of Upreti (5.04) than the Saharia of Taleti. It indicates that ever married Saharia women may expect 4.2 children in her life-time. Saharia indicate an IMR of 85.9 which is higher than Rajasthan (83.0) and Indian national average (70). Saharia of Taleti show much higher IMR (90) than Saharia of Upreti.

Morbidity Pattern/Burden of Diseases

It has been found that majority of cases belong to tuberculosis (60.8%), skin diseases (scabies & eczema) (14.0%), malaria (3.5%), typhoid (4.0%), eye diseases (4.0%), mental illness (3.0%), diarrhea (3.0%), asthma (2.0%), Sexually Transmitted Diseases (STDs) (2.5%) and others (3.0%) are diseases that occur regularly among Saharia.

HEALTH SEEKING BEHAVIOUR

Concept of Health

Saharia believe in supernatural powers and they have strong feeling that different deities and spirits are responsible for causation of different diseases. The spirits of dead ancestors are also believed to play an important role in ensuring good health, prosperity and protection to the community. Among Saharia, there is a strong belief that persons emanating evil, having mystical powers like evil eye and evil touch can also affect or deteriorate the health of a person. There is a traditional healer ('Janteyar') in the community who identify the cause of illness and treatment is based on it. The traditional healer ('Janteyar') develops faith and assurance for good health in the patient and treats him. Physical isolation, ecological legacy, traditional culture and beliefs have made the Saharia self-reliant and they have developed protective mechanism for their survival. Majority of Saharia especially in the interior villages are ignorant about the modern amenities and still follow the rituals of pleasing supernatural powers.

Concept of Illness

When a person could not move, is bed ridden, has physical discomfort, has loss of appetite, is unable to peform daily activities, is having certain physiological disorder, a person is considered to be ill by majority of the Saharia. A man or woman is considered sick if he or she is not able to do physical labour and withdraws himself from the work assigned. Saharia also believe that certain diseases take time to reveal physical mal-functioning.

SOCIAL/ENVIRONMENTAL CONDITIONS

In the past, food habits of the Saharia were fully wild and primitive. Various forest products and flesh of different animals formed their food. They were carnivorous. Among the forest products, which the Saharia used in the form of food include mahua, chirongi, gum, honey and amwala. The grains which Saharia used to consume in the past were of inferior quality from the point of view of nutrition. Now the Saharia eat Jawar, Bajra and Wheat. But the consumption of cereals, pulses and vegetables is very less, mainly because of low production and the cost factors.

Houses of Saharia do not have any arrangements for light and air in the villages. Poor lighting during night hours, absence of suitable exit for smoke, i.e. too much smoke in crowded and ill-ventilated house may be the cause for the high incidence of 'ratondi' (night blindness). Non-availability of fresh air during night causes respiratory troubles. The traditional customs, ignorance (illiteracy) and poverty, on the one hand and indifference and reluctance of the management in these regards on the other hand are the main causes for their living in this condition.

Sanitation among the Saharia is very poor. It has been observed that the Saharia go to the field for defecation. The existing habits of defecation may be responsible for ill health of Saharia in those areas.

Personal hygiene is also quite poor among the Saharia. Due to poverty, illiteracy and poor awareness about the importance of personal hygiene, the Saharia are unaware and ignorant about the fatality of such poor environmental sanitation.

Many infectious diseases such as cholera, typhoid and polio are carried in water. The principle source of these infections is the pollute drinking water from village ponds, rivers, hand pumps, tanks etc. The water is taken directly from rivers, wells and hand pumps for consumption. During summer the hand pumps and wells get dried and water is collected in water tanks. Water crisis in the interior villages becomes acute during summer.

The Saharia are fond of smoking and drinking. They use tobacco and locally available 'Tendu' leaves (Diospyros melanoxylon) for smoking in the form of 'Bidi' while locally prepared alcoholic drink using some forest fruits and 'Mahua' flowers (Madhuca indica) is much popular, there is no restriction on females either for smoking or drinking along with their spouses.

PATTERNS OF TREATMENT

Ethnomedical Treatment

In the tribal society, the social and cultural activities revolve around gods and spirits. They have specific gods for each and every activity of life. Likewise, they have gods for health, disease and calamities. They worship these deities to protect them from evil spirits, diseases and epidemics. For example the Saharia worship Tejaji, a deity related with snakes. Tejaji protect them from snakes and save their life in forest.

Like other tribes in India the Saharia, have also erected structures, though not as elaborate, conspicuous and posing as found among the non-tribal rural communities to house the images of various kinds of supernatural beings. Both at the individual as well as at the community level, the Saharia realise the responsibility of a safe placement of such images. Represented through various symbols and figures of various forms and material, they are kept at one or different places in and around the tribal settlements or in the near proximity of natural surroundings. Such religious places remain open to one and all irrespective of age, sex, etc. Man's interaction with such supernaturals does not merely end in the preservation of symbols and images. Rather, it is seen that they are worshipped and appeased on specific occasions in the year. Offerings of various kinds, sacrifice etc. are made to them on specific occasions. The will of community holds strong to keep up such values and traditions.

The Saharia believe that these supernaturals if disturbed through any gesture or forgetfulness, can cause severe condition that would prove damaging to the community health. The Saharia try to appease these supernaturals by offering prayers, worshipping them regularly and also on specific occasions like religious festivals, marriage ceremonies and rituals etc. The Saharia also have their own explanations of sickness, on the basis of which they developed their peculiar ways of curing diseases. This ethnomedical system (traditional medical system) is still in practice to a considerable extent especially among the Saharia who live in interior areas.

The Saharia believe that diseases such as measles (badi mata), chicken pox (sitla mata), unsafe delivery, snake bite, fever, typhoid (bade maharaj), malaria (chotte maharaj), pneumonia (dibbirog), tetanus (dhanusban), fits (dakani lagna or bhoot lagna). There is hierarchy of deties and some dities are placed in senior positions. A 'Janteyar' is consulted for treatment these diseases. The 'Janteyar' also checks the 'Nabj' (Pulse) of the patient and tells to make offerings of coconut, ghee (clarified butter) and sugar wafers (patasha) on the platform of deities. He also gives spiritual ash to the patients. Incense sticks are also lighted to please these gods. Consumption of liquor, meat and oil in the vegetables is avoided during days of sickness. The socio-cultural tradition does play an important role in the context of health and treatment among the Saharia. The common beliefs, customs and practices connected with health and diseases have been found to be intimately connected with the treatment of disease. When a person goes to traditional medical practitioner he gets socio-psychological reinforcement which he cannot get from the modern medical practitioner. The social reinforcement based on the understanding of traditional beliefs and practices is often lacking and it may be one of the reasons for making a modern medical practitioner unsuccessful in the Saharia areas. For some ailments, the Saharia resort to herbal medicine which they prepare themselves with the help of various plants found in the forest. It has been observed that health and treatment are very much connected with the forest ecology. The traditional health care system and treatment are based on their deep observation and understanding of nature and environment. Various plants have been used by the Saharia for medicinal purpose. Indigenous medicine is used by the Saharia for curing certain ailments. The Saharia believe in taboos and totems. The totem was ascribed a guardianly role to the clan and commemorated with rituals and worship. Rigid restrictions and avoidances came to be associated with the totem i.e. the totemic taboos. The Saharia belonging to Parondiya and Kaluj clans regard Banyan tree as their totem. Saharia (belonging to Parondiya and Kaluj clans) do not cut Banyan tree and cutting of Banyan tree is considered a taboo. Saharia belonging to Raichohan, Solvia and Garwar clans consider Asapal tree as their totem. The totems are regarded by the Saharia with awe and fear. The occurance of diseases are contributed to their wrath. The Saharia are aware of the mechanism of warding off the dangers resulting from the violation of a taboo. This involves the acts of penance and ceremonies of purification. For this purpose they recite the name of the totem, burn the incense and offer sweets to please the deity and later the sweets are distributed to the children. The totems are held in high esteem by every Saharia family and usually images of the totem are depicted either on paper or on walls and are worshipped every morning and evening by the elders of the family. The other occasions on which totems are worshipped are the death and marriage ceremony. The blessings of the totem are sought for the departed soul to find a place in heaven. In the marriage ceremony, image of the totem made of the flour, or drawn on the paper or walls is worshipped at every stage. After the marriage, the bride and the bridegroom pay their tributes to the totem by offering rice and vermilion. During the Holi, Deepawali and Navratra festivals the Saharia worship their totems along with other Hindu Gods. The 'Janteyar' prescribe measures for protection from evil eye. For cure of the problems the 'Janteyar' enjoy immense faith in the Saharia society. The 'Janteyar' also ties amulets ('Tabij') around the neck and forearm of children to protect them from evil-eye. He also gives 'Jhadà' treatment to a victim of psychic problems particularly fits.

The Saharia use varieties of roots, shoots, leaves, flowers, barks, seeds of forest, trees and bones and flesh of animals and birds available in the forest for the purpose of medicines and cure of diseases. These indigenous medicines are used by Saharia for curing diseases such as Bukhar (Fever), Saradi-Khansi (Cough and Cold), Sirdard (Headache), Deha Darda (Bodyache), Katana (Cut), Jalana (Burning), Ghaw (Wounds), Khujali (Itching), Ulti (Vomitting), Dasta (Loose motion), Petdard (Stomachache), Anakha Aana (Infection of eye), Bhukha Appetite), Pathari (Stone), Pilia (Jaundice), Chinia (Sugar), Dhata (Night fall), Safed Paani (Leucorhoea-White Discharge) and Kabj (Constipation).

Allopathic Treatment

In addition to the magico-religious method of treatment the Saharia also follow allopathic system of medicine. In the past the Saharia always used to consult traditional healers for treatment of sickness. But now, with the advancement in science and technology faith in supernatural concept is on decline and the Saharia have faith in allopathic doctors also. But the Saharia who live in interior areas mainly practice ethnomedical system. However it was observed that in most villages (75%) different treatments (spiritual, herbal and allopathic) go side by side and people have equal faith in all the treatments. They don't give preference to any particular method of treatment. The services of qualified doctors are inaccessible to a majority of Saharia, especially in remote villages and where these are accessible the poor Saharia patients are neglected by the doctors. In such circumstances, they prefer to go directly to nearby private unqualified doctors where good treatment at low cost is available.

61.3 per cent Saharia of Upreti and 58.7 per cent Saharia of Taleti have ever utilized the PHC services. It has been found that out of a population of 5675 among Saharia of Taleti (8 villages) there were 1435 (25.2%) sickness cases whereas in Upreti (8 villages) the number of sickness cases were 816 (22.8 %) as reported by the medical doctor. The data was collected from 104 sickness cases in Taleti and 95 sickness cases in Upreti. Out of 104 sickness cases in Taleti 9.6 per cent patients were taking herbal treatment, 29.8 per cent patients were taking treatment from traditional healer ('Janteyar') treatment, 52.9 per cent were consulting PHC and 7.7 per cent were not taking any treatment. Out of 95 sickness cases in Upreti 10.5 per cent patients were taking herbal treatment, 24.2 per cent patients were taking treatment from traditional healer ('Janteyar'), 43.1 per cent were consulting PHC and 22.2 per cent were not taking any treatment. The percentage of Saharia (Taleti) taking treatment from PHC for various diseases is: Tubeculosis (40.3%), Skin Diseases (scabies & eczema)(3.8%), Malaria (0.96%), Asthma (1.9%), Typhoid (0.96%),Nightblindness (0.96%), Lecorrhoea (0.96%), Fits (0.96%),Accident (0.96%).

The percentage of Saharia (Upreti) taking treatment from PHC for various diseases is: Tuberculosis (37.8%), Skin diseases (3.2%), Asthma (1%), Accident (1%). However it was observed that in most villages (75%) different treatments (spiritual, herbal and allopathic) go side by side and people have equal faith in all the treatments. They don't give preference to any particular method of treatment. According to NGO workers of CECOEDECON, the Saharia are not able to purchase medicines due to poverty. Even if they could purchase they cannot continue them for a long period and this results in discontinuance of treatment after a short period. They also change doctors quite often during treatment. At an advanced stage of the disease many a times doctors also fail to treat it. Then the Saharia again have to resort to their traditional methods of treatment or continue to move from one doctor to another and spend a substantial amount of money on treatment. Continuity of sickness for a long time is not uncommon among the Saharia. The longer the period of sickness the higher is the expenses on treatment. According to Doctors of PHC Shahabad the most common ailments for which the Saharia approach the PHC are skin diseases like scabies and eczema, loose motions, malaria, infection of upper respiratory tract, pneumonia, injuries, eye infections and ailments due to malnutrition. Few Saharia have more faith in private practitioner (Unqualified) and prefer to spend money on him rather than going to the government hospital or PHC. The doctors in the PHC Sahahabad also told that the Saharia have more faith in injections than in tablets. They want immediate recovery and think it can be provided by an injection. A Saharia would take injections prescribed to him but has a callous attitude towards Tablets. However in the interior villages it was found that when a person falls ill initially he consults a spiritual doctor, locally called 'Janteyar'. When the treatment of 'Janteyar' did not give any relief and the disease enters it's critical stage then the patient consult an allopathic doctor. The new generation is changing with the modern times with the advancement in science and technology. But the Saharia of older generation still adhere to ethnomedical practices of curing diseases. It is only when the 'Janteyar' fails to cure the patient and his condition becomes bad to worse then he is taken to the allopathic doctor for treatment. The incidence of Tuberculosis (60%) is very high among the Saharia. This is mainly because of their low living standard, bad personal hygiene, smoking and drinking habits. The incidence of skin diseases like scabies ('Dad') and eczema ('Khujali') is also high (11.5% in Taleti and 17% in Upreti) due to unhygienic living conditions. Other diseases prevalent among Saharia were Nightblindness ('Ratondi'), Malaria ('Chotte Maharaj'), Fits ('Dakini Lagna' or 'Bhoot Lagna' or 'Upri Hawa'), Typhoid ('Bade Maharaj') Diarrhoea ('Ulti-Dast'), Stones ('Pathar'). During the fieldwork it was found that for snake bite cases Saharia always consulted 'Janteyar' because they think that doctors are helpless in curing snake bites and they can do anything .The Saharia of Taleti (60%) and Upreti (61%) who were suffering from TB (age ranges from 30-55 years) were found attending TB Control Camp which is held on 23rd of every month at Shahabad. It shows that the new generation is changing with modern times. The Saharia have identified some diseases (fits, snakebite) for which they mostly consult 'Janteyar'. If the 'Janteyar' fails to cure the patient only then they go to PHC for treatment. However in case of accident they consult an allopathic doctor for getting treatment at the PHC because they think that 'Janteyar' cannot cure it.

Case studies revealed that Saharia consult 'Janteyar' for treatment of diseases. They believe that diseases are caused by wrath of deities. They use indigenous medicines also for treatment of diseases. Though most of the Saharia have visited the PHC but the Saharia are not happy with the services of PHC. They mentioned that at PHC the doctor gives prescription on paper only and the patients were not given any medicine at PHC. Sometimes the doctors refuse to see the patient then they consult Private doctors who pay attention on them. They are not happy with the behaviour of the PHC staff. The new generation is changing with modern times but the Saharia of older generation still adhere to ethnomedical practices of curing diseases.

Maternal and Child Health Care

No special care is taken of a pregnant woman among the Saharia. She takes the routine diet during the entire period of pregnancy. Milk is given only if the family has milching cattle. The pregnant woman works in field also. The Saharia women know that intake of milk, dalia, green leafy vegetables is good for the growth of foetus. But due to poverty they cannot afford to eat what they should eat during pregnancy.

Focus Group Discussion (FGD)

Two FGD were carried out among the Saharia of Taleti. Eight male individuals were selected in each FGD, who belonged to middle age group (30-40 years). All the participants were illiterate. Their villages (Pathari & Sandhri) were not connected with pucca road. Components taken into consideration for FGD were; concept of illness, health seeking behaviour, attitude towards traditional healer and felt need of tribal population with regard to health services. 75 per cent participants from Pathari and 87.5 per cent participants from Sandhri reported that whenever they felt physically uncomfortable they considered themselves sick and remaining participants from both villages reported that when they could not carry out daily routine work they considered themselves sick.All the participants (100%) reported that there was no doctor near their village whom they could approach for seeking treatment. All the participants (100%) revealed that two important factors which influence treatment were firstly severity of ailment and secondly proximity of the doctor. All the respondents (100%) revealed that for minor illness they preferred traditional healer who resides in their village. But if the illness got severe then they go to Sipri (M.P.) 75 per cent respondents from Pathari and all the respondents from Sandhri mentioned that when they go to PHC which is at a distance of 30 Kms from their village, the doctors did not give them much attention and medicine were also not available at PHC so they don't prefer to go to PHC. 62.5 per cent respondents from Pathari and 75 per cent from Sandhri reported that whenever any one fell sick, after waiting for one or two days, the patient was taken to private practitioners who reside near their village. For minor ailments traditional healer ('Janteyar') is consulted. All the respondents (100%) from both the villages revealed that private doctors were preferred for prompt attention and quick relief. All the respondents reported that private doctor gave more attention to patients. 50 per cent respondents from Pathari and 50 per cent from Sandhari also reported that private doctors used stethoscope, checked pulse of the patient and also listened to patient's problems. 62.5 per cent respondents from Pathari and 75 per cent from Sandhri also revealed that when the condition of patient becomes very serious they were taken to either block PHC or Hospital at Sipri (M.P.). In fact during serious condition private doctors themselves referred the patient to these places. As revealed by 87.5 per cent participants from Pathari and all the respondents from Sandhari said that the transportation system was very poor and patients found it difficult to travel by crowded buses. 62.5 per cent respondents from Pathari and per cent 75 per cent from Sandhri revealed that doctors at PHC did not spend much time with patient's problem and also said that medicines were not available in BPHC.

After waiting for long time, doctor wrote the prescription and medicine was supposed to be purchased from the chemist shop at Shahabad. Doctors were also in hurry as a result it became difficult for the patients to understand his advice. 87.5 per cent respondents from Pathari and 87.5 per cent from Sandhri reported that traditional healers ('Janteyar) were preferred, in order to eradicate the evil power from person whenever child cries continuously or if the person was sick for a long time, In such cases the help of 'Janteyar' was sought. These respondents also revealed that the help of 'Janteyar' was needed when a person is bitten by snake.

Two FGD were carried out among the Saharia of Upreti. Eight male individuals were selected in each FGD who belonged to middle age group (30-40 years). In the village Hatri, out of eight individuals, four had completed their primary education, two had completed secondary school education and remaining two were illiterate in the village Ghesua. PHC is at a distance of 20 km from the village Hatri. In village Ghesua eight males aged (30-35) were selected for interviews. None of them was literate. Same components were taken into consideration for Saharia of Upreti. The PHC is at a distance of 9 Km from Ghesua. 37.5 per cent participants from Ghesua and 50 per cent participants from Hatri reported that whenever they felt physically uncomfortable they considered themselves sick. Rest of the participants from both the villages reported they considered themselves sick only when they couldn't get up from bed, unable to comply with daily routine work. 65 per cent participants from Ghesua and 50 per cent participants from Hatri reported that cold and cough, bodyache, low temperature fever, diarrhoea and skin diseases were not considered unusual, in other words, individual showing any of these symptoms were not considered as sick. Since they were very poor, they couldn't afford to continue the treatment for long time and also were unable to buy medicines for every small cause. 75 per cent participants from Ghesua and all the participants (100%) from Hatri reported that when there was nothing to eat at home, expenditure on health became a secondary factor except when the person was very sick. 62.5 per cent participants from Ghesua and 37.5 per cent participants from Hatri reported that they used to go to Primary Health Centre (PHC) whenever they were sick. All the participants from both the villages felt that choice of specialist depended upon seriousness of illness. Cost for the treatment was an important factor to decide whether they would seek treatment or not was felt by 75 per cent participants from Ghesua and 65 per cent participants from Hatri. 62.5 per cent participants from Ghesua complained that compounders and nurses were very rude and they treated poor with great degree of indifference. They also reported that medicine was not available at PHC or Sub-Centre and patients were asked to buy medicine from chemist shop at Shahabad. However, the rest 37.5 per cent participants reported that medicines were available at PHC but the nurse or compounder did not give medicines to them. 87.5 per cent participants of Ghesua reported that the help of traditional healer ("Janteyar') was sought in cases like snake bite and fits etc. However all the participants (100%) of Hatri reported that cases of snake bite and fits were cured by 'Janteyar'.

Two focus group interviews were carried out among the Saharia of Upreti with regard to Maternal and Child Health Care (MCH). Interviews were carried out in two Upreti villages (Khandasehrol and Dikhwani). FGD was carried out among a group of six women in each village. The interview was carried out among six ever married women in the age group of 25-30 years from each of the villages. None of the women from these villages completed their primary school education. Components taken into consideration for FGD were; awareness about ANC check-ups; factors restraining women from utilising ANC services, consumption of food during pregnancy and child immunization. 59 per cent of the participants related ANC with the consumption of iron and folic acid tablets and tetanus toxoid injections. 41 per cent of the participants expressed their unawareness. 51 per cent of the participants felt that pregnancy was a physiological event and did not require any medical care. 50 per cent participants from these villages felt that their mother-in-laws do not allow them for medical check-up. The mother-in-law feared that such check-up might harm the child. It was also reported that their mother-in-laws rely more on elderly ladies and traditional dais. They also expressed that they could not rely on anganwadi workers and health workers for ANC check-up, as they are young and have little knowledge and skill regarding birth of a child. Communication and transport were the major problems expressed by 40 per cent respondents. 50 per cent participants reported that it is not possible for pregnant women to go for regular check-ups. The bullock-carts were often the sole mode of conveyance between the peripheral village and the nearest health facility. It took long time and hence difficult for women to travel by these means. Private vehicle became too expensive for regular check-up during pregnancy. 60 per cent participants reported that the state Government buses were operating from Baran to Sipri but these buses are over crowded. 28 per cent participants reported that no health personnel visited their village for distribution of iron tablets or for immunizing pregnant women. All the participants reported that usual food was consumed during pregnancy. No special care is taken of pregnant women among the Saharia. She takes the routine diet during the entire period of pregnancy. Milk is given only if they have milching cattle. 45 per cent of the participants from these villages of Upreti could tell utility of each specific vaccination and number of vaccinations. However 60 per cent participants from the villages revealed that polio vaccine saves the child from physical handicap.

Two FGD were carried out among two villages (Beelkhera Dang and Faredua) of Taleti. FGD was carried out among a group of six women in each of the two villages. None of the women were literate. Components taken into consideration for FGD were same that of the Saharia of Upreti. 58 per cent of the participants expressed their unawareness regarding ANC check-up. Only 42 per cent were aware of it. 66 per cent of the participants felt that there was no need for ANC check-up during pregnancy since it was believed that pregnancy was a natural process. 84 per cent participants reported that since the Block PHC was 15-20 km. away from the village, it was not possible for them to go for regular check-up. It was also felt by respondents that they were not aware of the time, when health worker visited their village, since they used to go out for work early in the morning. To avail such services, would result in loss of their daily wage. 68 per cent participants also told that someone had to accompany pregnant women, this would result in loss of wages for both of them. BPHC was always over-crowded and women had to spend 2-3 hours at BPHC for different check-ups from one person to another and it became very tiring. All the participants from both these villages maintained that usual food was consumed during pregnancy. No special care is taken of a pregnant woman among the Saharia. Saharia woman takes the routine diet during entire period of pregnancy. 82 per cent women felt that instead of such preventive measures if food or money were provided to them to meet their needs, it would have been more beneficial. 32 per cent women participants from these villages could name the different vaccination, number of doses or their utility. However 42 per cent women felt that although they were aware that immunization was good but they were not aware about exact time and the number of doses of vaccination.

Two FGD were conducted among two villages of Upreti (Samrania and Mahodara). Each FGD consisted of six married women in the age group of 25-35 years. Two women were having education up to middle school. Components taken into consideration in FGD were awareness of FP, sources of awareness, attitude towards FP, centre for adopting FP methods and felt need in FP. Sterilization was most widely known method among respondents. Knowledge regarding reversible methods was limited. 48 per cent of participants were familiar with the terms IUD, Copper T and Oral Pills. 52 per cent participants were aware of the person who provided traditional medicine for birth control, but could not give much information regarding the kind of medicine given. NGO's like CECODECON (Centre for Community Economics and Development Consultants Society), LOK JUMBISH PARIYOJNA, SANKALP were the major sources for spreading FP information among both the Saharia of Taleti and Upreti. PHC workers have also played a role in spreading FP messages. Inter-personal Communication has also played an important role in spreading FP information. Mass media had limited impact in spreading FP messages as expressed by respondents. However 34 per cent respondents had seen pamphlets and posters during FP camps. 62 per cent respondents from these villages reported that health workers did not give much information about reversible methods of FP. Much stress was given on sterilization. The respondents felt that the health workers were interested merely in knowing whether the couple had undergone sterlization or not. Information regarding reversible methods provided by health workers was not satisfactory as revealed by 82 per cent of respondents. They also expressed their own fear with the use of reversible methods, participants revealed health personnel deliberately did not tell the side-effects of FP methods, fearing that nobody would then opt for FP. 89 per cent of respondents reported that they were not in favour of FP because they felt that after adopting the FP the mothers become very weak and therefore, were not able to look after the family properly. They also feared that if something ill happened to their existing children, they would not be able to produce any more children in future, after the operation. 61 per cent respondents favoured FP, as they felt that due to their poor economic condition they would not be able to feed a large family. 98 per cent respondents preferred PHC as the place for adopting FP.

Two FGD were carried out among two villages of Taleti (Shahabad and Chaurakhadi). Each FGD consisted of six married women in the age group of 25-35 years. All the participants were illiterate. Same components were taken into consideration. All the respondents were aware of sterilization. However only 5 per cent participants were aware regarding Oral pills, copper T, IUD. 64 per cent participants favoured FP mainly due to their poor living condition and their inability to support large family. 88 per cent respondents preferred PHC as the place for adopting FP.

Impact of Health Programmes on Saharia

In the Shahabad and Kishanganj tehsils, various programmes related to health and medical care have been running in the Primary Health Centres, Sub-Centres and dispensaries. These programmes are: Primary healthcare, rural primary healthcare infrastructure, health manpower, control of communicable diseases, National Malaria Eradication Programme, Multipurpose Camps (for ENT diseases, treatment for Handicapped persons, Dental Check-up), National Tuberculosis Control Programme, National Leprosy Eradication Programme, Control of Non-Communicable Diseases, National blindness Control

Programme, etc. Some of these programmes are run by the State and Central Government).

It was observed that Tuberculosis control camp is organised on 19th of every month at Kelwara and 23rd of every month at Shahabad. Many Saharia who are suffering from TB attend this camp. Saharia have been given 'Chainit Parivar' cards when they feel ill they can use this cards and get medicines from the PHC. But when interviewed many Saharia told us that when they go to PHC the sister or compounder ask Rs.2/- per tablet for cold and headache and if they don't give money to them then the compounder or nurse scolds them and don't give medicines. The Saharia are not afraid of immunization. When PHC Staff visit their colonies then they get their children immunized. All the children below one year were found immunized at the time of fieldwork.

82 per cent women of Taleti and 92 per cent of Upreti were knowledgeable about sterilization as a method of FP. 2.7 per cent women of Taleti and 2.0 per cent of Upreti were aware about oral pill, only 2 per cent women of Taleti were aware about IUD, 10.6 per cent women of Taleti and 6.0 per cent of Upreti were aware about traditional method of FP and 2.7 per cent women of Taleti were aware about Multiple choice (more than one method of FP). Thus, the knowledge about FP method among the Saharia women of Taleti and Upreti is mainly restricted to terminal method of FP. 58.7 per cent Saharia women from Taleti and 52.0 per cent Saharia women from Upreti had never seen any material on FP. Further 25.3 per cent women from Taleti and 32.0 per cent from Upreti had seen film show on FP. 9.3 per cent women (Taleti) and 8.0 per cent (Upreti) have seen TV/Radio on FP. 52.0 per cent respondents from Taleti and 54.0 per cent respondents from Upreti were reported to have come to know about FP from PHC/anganwadi workers. Another 24.0 per cent

(Taleti) and 20 per cent (Upreti) have come to know about FP through interpersonal communication (Friends/ Neighbours), 8.0 per cent women of Taleti and 12.0 per cent of Upreti have come to know about FP from Private Practitioners.

It was observed that 64 per cent Saharia of Taleti and 86 per cent Saharia of Upreti wanted to restrict the number of children. 16 per cent Saharia (Taleti) and 12 per cent Saharia (Upreti) mention that it was the mother-in-law who were decision maker. However 8 per cent Saharia (Taleti) and 16 per cent Saharia (Upreti) revealed that the decision for adopting FP method was the joint decision of both husband and wife. Another 32 per cent Saharia (Taleti) and 22 per cent Saharia (Upreti) stated that husband was instrumental in the decision regarding the use of FP method. It was observed during the discussions that the need for family planning is widely felt by the tribal women, though the men, by and large, are not very keen on it. To begin with, during community discussions, men folk maintained that it was the women who had the final say on family planning matters; but the women asserted that though they played an important role in FP matters, it was their husband's who took the final decisions on FP matters. The responses of those who have ever used any FP measures and who have not used were taken into consideration. Poor economy was the reason for 32 per cent Saharia (Taleti) and 62 per cent Saharia (Upreti). 28 per cent Saharia (Taleti) and 20 per cent Saharia (Upreti) wanted to go for FP because then wanted to have small family. 8 per cent Saharia (Taleti) and 6 per cent Saharia (Upreti) considered motivation as influencing factor for adopting FP method. It was found that 70 per cent Saharia (Taleti) and 88 per cent Saharia (Upreti) visit PHC to avail FP services. 12 per cent Saharia (Taleti) and 6 per cent Saharia (Upreti) preferred traditional practitioners (Janteyar) for availing FP services. 6 per cent Saharia (Taleti) and 2 per cent Saharia (Upreti) go to private practitioners for availing FP Services.

Saharia Development Project (SDP)

Saharia Development Programme was launched in 1977 under the chairmanship of District Collector. There are a number of schemes through which efforts are being made for the socio-economic upliftment of the Saharia under SDP. The project is being financed mainly through Central assistance (94.33%) and 5.67 per cent from state plan. The various heads of the project include agriculture, irrigation, rehabilitation, rural housing, electrification, animal husbandary, forest development, medical and public health, drinking water, training, education, cooperation, free legal advice, road construction, other development programmes and administration. Implementation of SDP is the sole responsibility of Saharia Development Committee. The Committee is headed by the District collector. Members of the parliament and legislative assembly from the area, zila pramukh, two representations of Saharia Community and related District level officers are members of the committee. Project Officer of the Saharia Development Project is the Additional Divisional Magistrate (A.D.M.). Funds have been provided under the SDP for free distribution of medicines to Saharia, establishment of Health subcentres, organisation of medical camps, family welfare programme and upgradation of PHC Shahabad into twenty-beded hospital, along with residential quarters for the staff. However it has been observed that the implementation of medical and health schemes has been very slow. In the PHC Shahabad medicines are not available. X-ray machine is there but it is not working.

Role of NGO's

In addition several voluntary organisations like ADIM JATI SEWAK SANGH, CECOEDECON, ASEFA and

SANKALP are also working for socio-economic upliftment of Saharia. They are trying to create awareness about education, medical and health facilities, safe drinking water, family planning etc.

Views of Health Providers

Medical Officers at PHC were mainly treating cases of malaria, diarrhoea, scabies, eczema, cough and fever. But it was observed that special stress was given on child immunization and family planning. Limited laboratory facilities were available at PHC. Although vehicle facility was being provided at Block Primary Health Centre, it was used only for bringing medicine from district headquarters. Most of the drugs prescribed to the patient had to be purchased by the patients from the chemist shop. However in the interior villages there is no chemist shop. The health providers felt that chief reason for poor health situation among the Saharia were lack of awareness, illiteracy, poverty, low socio-economic condition and bad sanitary conditions. Providers felt that most common diseases among Saharia were Tuberculosis, Scabies and Malaria. Providers felt that poor means of communication between villages and health centre led to under utilization of health services. They also felt that due to high price of drugs and poor economic condition tribal people could not continue their treatment for long and stopped taking treatment as soon as they felt little comfortable without completing the full course of medicine. As a result their health condition did not improve. They approached health centre only when the condition of patients became worse. They blamed the traditional healers for under utilization of health centre by tribals. They felt that since these traditional healers who reside in tribal villages, were better known among tribals. Tribals approached these healers and Bengali doctors (unqualified doctors) more often as compared to PHC. Since these healers and unqualified doctors were not formally trained in allopathic medicine, often prescribed wrong medicine. It was also reported that in order to recover quickly, in some cases tribals observed to have taken overdose of antibiotic prescribed. It was reported by Block Supervisor PHC Shahabad that Block Primary Health Centre was ill equipped to meet the needs of thousands of patients. It was running short of medicine and manpower. Health providers strongly felt that in order to improve the health condition among tribals there was a need for increasing literacy among tribals. Health workers felt tribals were highly negligent and illiterate because of this they did not utilise health services properly. Besides this their poor unhygeinic living condition, drinking and smoking habits were also responsible for their poor health. Health workers expressed that it was very difficult to convince the tribals for utilizing preventive services. The health workers felt that utilization of ANC was poor among tribal women. Though they supposed to make four visits for ANC but it was told that it was not so. Health workers told that if tribal women immunized their child, most of them did not complete the full course of immunization. Health workers reported that there.was tremendous work pressure on them. The villages covered by them were scattered and sometimes when they visit the household and the mothers were not available. It was also reported that due to shortage of staff they had to do additional work besides their normal duties. Female health workers reported that during rainy season, it usually becomes very difficult for them to cover all the villages of their Sub-centre as the villages were cut-off from pucca roads.

All the Health Providers felt that illiteracy and unawareness were major hurdles for increasing their level of consciousness towards better healthcare. Health providers also suggested that by improving the communication channel between providers and tribal community better delivery of healthcare could be ensured.

Felt Needs

Focus Group Discussion carried out in the villages (Pathari and Sandhri) revealed that all the participant (100%) from both the villages felt that a hospital should be opened in the villages especially in the interior villages. There should be regular supply of medicines at PHC. All the participants said that there is a dire need for supply of medicines at PHC. Focus Group Discussion conducted in the villages (Ghesua and Hatri) revealed that all the participants (100%) of Ghesua and Hatri felt that there should be regular supply of medicines at PHC because due to their poor economic condition they cannot purchase medicines. They also felt that this was the main reason that treatment of a person was not continued. 50 per cent participants of Hatri reported that health workers do not visit the village regularly and there was no one to approach in case there is any emergency. 75 per cent participants from all these villages felt that proper house to stay was important for good health. All the participants (100%) of the villages felt that adequate amount of food was most important factor maintaining good health. 48 per cent respondents from the villages Samrania and Mahodara told proper information should be given to users regarding the side effects of contraceptive methods. The need of opening a hospital or dispensary was more keenly felt by the Saharia in the interior villages of Taleti especially Beelkheradang (55%), Chaurakhadi (50%), Hari Nagar (60%), Pathari (70%) and Sandhri (55%). On the whole 43.5 per cent Saharia of Taleti want opening of hospital or dispensary as compared to 27.5 per cent of Upreti, 20 per cent Saharia of Taleti as well as Upreti want a doctor to be posted in the villages, 26 per cent Saharia of Taleti and 18 per cent Saharia of Upreti want free and timely supply of medicines. However 4.5 per cent Saharia of Taleti and14.5 per cent Saharia of Upreti want no other facilities than the existing ones. 6 per cent Saharia of Taleti and 20 per cent Saharia of Upreti did not responded.

A Brief Comparative Account of Status of the Tribal Health

Tribals in India are distributed in varying concentration in different parts of the country. Each tribal community has its own socio-cultural practices, health practices, taboos and distinct identity of their own lineage and they stick to it. Their health status, food habits, life styles and diseases and their remedies may differ from each other. Anthropologists, social scientists and social workers in different parts of the country have studied health status of the different tribal communities. Studies on specific tribal community in selected tribal pockets have been conducted by various scholars (Dave, 1960; Opler, 1963; Basu, 1967; Hasan, 1967; Ghosh, 1970; Jay, 1970; Kumar and Mitra, 1975; Murthy and Ramesh, 1978; Basu, 1985, 1990, 1998; Banerjee, 1988; Dutta, 1990; Gurumurthy, 1990; Ray and Roth, 1991; Bhasin, 1994; Singh, 1994; Swain, 1994 and Kshariya, 1997).

The sex ratio among the Saharia of Upreti was found to be 903.2 and 897.1 among the Saharia of Taleti. Low sex ratio among the Saharia can be attributed to their poor utilization of health services, primitive delivery practices and inaccessibility to the health institutions for safe delivery. It has been observed that the sex ratio of Saharia (897.1) is lower than the sex ratio of Rajasthan (910) and sex ratio of India (933). The present study shows Crude birth rate of 44.3 in Saharia (Upreti) and 27.8 in Saharia (Taleti). Infant mortality rate among Saharia has been found to be 85.9, which is higher as compared to Infant mortality rate of Rajasthan (83.0) and India (70.0). The infant mortality rate among the Saharia (Upreti) has been found to be 82.9 and Saharia (Upreti) 85.9. Higher Crude birth rate among the Saharia (Taleti) shows that more number of Saharia (Taleti) are adopting family planning methods. Crude birth rate of Saharia (34.7) is the same as that of Rajasthan (34.7). General fertility rate among the Saharia of Upreti was found to be 168.3 and among Taleti was 113.2. The total fertility rate among the Saharia (Upreti) is 4.85 and 3.35 among the Saharia (Taleti). Age specific fertility rate among the Saharia (Upreti) is 0.97 and among the Saharia (Taleti) is 0.67.The higher fertility rate might be due to lesser number of Saharia (Upreti) were accepting family planning. Total fertility rate among Saharia is 8.2. Crude death rate among the Saharia (Upreti) is 10.4 and among Saharia (Taleti) is 11.9. Crude death rate among Saharia is 11.9, in Rajasthan is 10.4 and in India is 8.7. The higher crude death rate is due to lack of nutritious food, inadequate knowledge of health care, lack of accessibility to medical and health institution, social and cultural practices that are interwoven in their life and unawareness about preventive services.

Although the concept of well being and the notion of the disease varies between different tribal groups, living in different ecological zones with different cultures yet, in a tribal habitat, a person is usually considered to be afflicted with some disease if he/she is incapable of doing the routine work which is usually being expected to be carried out by that individual in the society, i.e., incapacitation from work is the universal index of poor health. Like other tribes in India the Saharia also believe in supernatural powers and they have strong feeling that different deities and spirits are responsible for causation of different diseases. The spirits of dead ancestors are also believed to play an important role in ensuring good health, prosperity and protection to the community. Among Saharia, there is a strong belief that persons emanating evil spirits, having mystical powers like evil eye and evil touch can also affect or deteriorate the health of a person. There are traditional healers ('Janteyar's) in the community who identify the cause of illness and treatment is based on it. The traditional healer ('Janteyar') develops faith and assurance for good health in the patient and treats him. They do not associate disease and epidemics with their insanitary condition of living but associated them to the wrath of the Gods and Goddesses. There is no concept of nutritious diet having good effect on health among the Saharia. Physical isolation, ecological legacy, traditional culture and beliefs have made the Saharia self-reliant and they have developed protective mechanism for their survival. Majority of Saharia especially in the interior villages are ignorant about the modern amenities and still follow the rituals of pleasing supernatural powers.

Few studies have been carried out to throw light on indigenous medical beliefs and practices and western medicine; treatment with religious rites; health services; impact of development programmes on tribal health; role of allopathic doctors in tribal areas; utilization of modern health practices and delivery system; pattern of health seeking behaviour (Singh et al, 1971; Jaggi, 1973; Banerjee, 1982; Rizvi, 1982; Gupta, 1986; Mann and Mann, 1986; Khan, 1986; Choudhary et al 1986; Banerjee, 1988; Bardhan et al, 1989; Ali, 1992; Basu et al, 1994; Sinha, 1994) whereas some of the studies (Roy Burman, 1990; Mahapatra, 1994; Bhattacharya, 1995; Ghoshmaulik, 1995; Kapoor, 1998; 1999) have recently touched upon the other dimensions of health.

Maternal and child care is an important aspect of health seeking behaviour. Maternal malnutrition leads to premature deliveries, low birth weight babies and pregnancy wastages. Few studies (Sharma,1979; Kapoor and Prasad,1982; Prasad and Kapoor,1982; Saha,1982; Rao et al,1986; Banerjee,1988; Basu,1990;93; Basu and Kshatriya,1989; NICD, 1993; Devarapalli, 1992; Prema and Thomas,1992; Pandey and Tiwari, 1993; Pandey et al,1996) have been carried out to study the maternal and child health care among different tribal groups of India.

Among the Saharia the old women of the family or neighbourhood help in the delivery. A 'Bhangan' (sweepress) is called to cut the umbilical cord ('Nal Katna'). The Saharia consider that it is only the work of the 'Bhangan' (sweepress). In the interior villages there is no sweepress and the sweepress is called from other village which requires two to three days and this has an adverse effect on the health of the mother and the newly born baby. The 'Bhagan' cuts the umbilical cord with blade and knife. The placenta is burried in the cattle shed.

The above mentioned studies carried out on different population groups of India show that health status of the tribal population in India is very poor. Tribals in India are distributed in varying concentration in different parts of the country. Each tribal community group has their own socio-cultural practices, health practices, taboos and distinct identity of their own lineage and they stick to it. Their health status, food habits, life styles and diseases and their remedies may differ from each other. Every tribe is living in a different ecological zone, has a different culture and has different health care practices. But the health status of the tribals living in different ecology with different cultures present a dismal picture. The present study shows that health status of Saharia is poor. Numerous scholars have tried to establish poor health status of tribals with the help of morbidity, mortality and health statistics. The widespread poverty, illiteracy, malnutrition, absence of safe drinking water and sanitary and living conditions, poor material and child health services, ineffective coverage of national health and nutritional services have been traced out as factors responsible for poor health status of tribals. The present study on Saharia have also traced out factors like widespread poverty, illiteracy, malnutrition, absence of safe drinking water and sanitary and living conditions, traditional health care practices, superstitions, totems and taboos, poor maternal and child health services, lack of health infrastructure, inaccessibility of modern medicine to tribals, ineffective coverage of national health and nutritional services which are responsible for dismal health conditions prevailing among the Saharia. The present study shows that there exists a relationship between ecology, culture and health. It can be said from the present study that poor health status of Saharia reflected in their health care practices is the outcome of their ecological, social and living conditions, illiteracy, ignorance, lack of transport facilities, under availability of potable water etc. and in appropriate medical and health opportunities.

Factors Responsible for Poor Health Status of Saharia

On the basis of intensive and extensive fieldwork during 1999-2000 among Saharia the following reasons have been found responsible for their poor health status:

A large number Saharia suffer from Tuberculosis (60%). This is due to lack of nutritious food, unhygienic living conditions and smoking and drinking habits and other environmental conditions. The food taken by Saharia is neither nutritious nor does it meet the calorie, protein and vitamin requirements of the body. They eat inferior kinds of grain like jowar and bajra. Milk is consumed only occasionally. It was the striking feature of the tribal people that there was absolutely no restriction on females either for smoking or drinking along with their spouses. The smoking and drinking habits result in malnourishment, loss of body weight and decreased resistance to infections. Unhygienic living conditions, delayed treatment and malnutrition affect their health badly. Even pregnant and lactating mothers and children are not provided with nutritious diet.

It has been observed that poor economic condition (90%) is also responsible for low health status of Saharia. Due to poor economic condition the Saharia are not able to buy milk and vegetables.

Several diseases are caused due to lack of safe drinking water. In some interior villages hand pumps are not working. Pond water is used for drinking as well as for washing and bathing. The Saharia, as a whole, do not attach much importance to sanitation aspects, be it sanitary latrines or systematic disposal of waste products.

Superstition is also a factor responsible for poor health status of Saharia. In case of fits when a person falls ill initially he resort to magico-religious treatment and if the sickness continues and disease enters its critical stage only then he consults an allopathic doctor. During pregnancy the Saharia do not consult a Govt. Doctor in order to have MCH services. They approach the Govt. Doctor only in case of complication or abnormality. Delivery conditions in interior villages were found to be horribly unhealthy. There is no local sweepress in some interior villages for cutting the umbilical cord (Nal Katna). When a delivery is conducted, she is called from other villages and it takes 2 to 3 days. This also affects mother and child's health.

Due to Kaccha roads and lack of transport facilities it is not possible for health workers to reach the interior villages to perform their duties adequately. In rainy season i.e. July-Sept it is not possible to approach villages in the interior and also not possible to conduct fieldwork especially in the villages Pathari, Sandhri and Chaura Khadi. During rainy reason as the area is not approachable even by jeep because the tyres of jeep get trapped in the clay and slushy soil and that is why the village becomes unapproachable. The Saharia live in remote and inaccessible and forest area which makes the health worker more difficult to frequently interact with the tribal. Although they cover the smaller population than his counterpart in rural area, it is not possible for the former to cover the entire population under his jurisdiction because of inaccessibility and lack of communication facilities. This goes to suggest that the size of population to be covered by health worker is important but more important is to see whether the health worker in tribal area can have an easy access to the area that he is supposed to cover.

Even the post and telegraph facilities are non-existent in the interior villages. In case any epidemic breaks out there would be massive peril before the news reaches the Block Development Officer (BDO), Additional Divisional Magistrate (ADM) or Sub Divisional Officer (SDO). The way to reach the news of such epidemic to relevant authorities is to go to them on foot for a substantial or by bicycle to cover the whole distance.

Lack of education is another factor responsible for poor health of Saharia. The uneducated people believe that more children means more working hands and they do not adopt family planning measures. Only women were found adopting family planning measures. They also have strong faith in magico-religious treatment of curing illness. Due to lack of adequate health education activities the Saharia are still ignorant about the various types of diseases occuring among them, and their causes. The knowledge of positive health, hygiene and preventive measures to be adopted to avoid sickness are more or less absent in them.

There is no effort by the government to impart health education to the Saharia.

It was also found that some doctors and ANM are taking their duties and responsibilities lightly. They consider their posting to these tribal villages as a punishment. Some young doctors are preparing for Post Graduate (MD OR MS) entrance exams. Corruption and exploitation of different nature are also rampant.

The various development programmes, which include provision of health care services, development of health infrastructure have been imposed on Saharia from above rather than considering the heath needs and their limitations and realities. There is extreme lack of concerted effort to observe and evaluate various health programmes launched for the development of Saharia. The health infrastructure facilities provided in the PHC and Sub-Centres are not adequately equipped according to the health need of the Saharia. There is only one Primary Health Centre (PHC) at Shahabad. Neither there is any chemist's shop in the interior villages nor any doctor or health personnel to take care of poor tribals. Poor socio-economic status, physical distance, unawareness, ignorance, fear of exploitation and absence of lady doctor (gynaecologist) also pose major problems for the Saharia to avail modern health facilities at the PHC and Sub-Centres.

It has been observed that Saharia are in a state of transition. It has been found that the concept that tribals are resistant to modern medical care system is not true. If the modern health care facilities are available then they accept it.

Ecology plays an important role in shaping culture which plays an important role in creating structures of health. It has been found that Saharia live in a particular ecology with a particular culture. It has been observed that Saharia have their own traditional skills of exploitation of ecology and they create differential structures of health keeping in mind the various socio-cultural practices.

It is suggested that it is necessary for health functionaries to have knowledge about the culture and society of the tribal people. The inadequate nature of facilities in the tribal areas, lack of respect for indigenous culture, lack of proper facilities and attention towards the patients are mainly responsible for the distrust of tribal people towards modern medicine.

Since the concept of health occupy different meaning in different tribes living in different ecological zones with different cultures therefore the health parameters of a community can not be studied in isolation because health is deeply interwoven with ecology and culture.

It has been observed that Saharia living closest to nature are influenced by the socio-cultural and environmental dimensions in their health care practices. They satisfy their immediate needs from the nature. Thus, environment – both physical and social, play an important role in the formation and development of a culture. Physical environment includes geographical location, climate, flora and fauna, etc. Socio-cultural environment of a people results mainly from the attempt of the group to adjust with their physical environment. It includes all the activities of a group of people right from food habits, marriage pattern to religious beliefs, etc. A change in environment may slightly affect the culture but in no case it can destroy the culture as a whole. Besides allopathic and spiritual treatment, herbal or indigenous medicine is their other preference for getting treatment which is locally available in the forest. Man is a cultural animal and it is not possible to discuss any human problem without social and cultural factors. "Since culture is prior to the person, and indeed the organism becomes a person principally as a direct effect of the cultural stimulation and organization, which has inherent potentialities and his acquired behaviour undergo, it becomes of the first importance to understand the nature of culture or cultures in which the person was culturalized, if one is to understand the person"(Montagu, 1963).

The herbs and plants required for treatment are collected from the nearby forest. Health is a common theme in all cultures. As health is a major aspect of culture, impact of environment on it can be easily traced. Forests are the inseparable part of the lives of the Saharia. In their world-view, day-to-day life and all round life activities, forests have a very important role. Health Care Practices of Saharia are greatly dependent on the forests and forest products. Their socio-psychological and cultural needs are fulfilled by the forests and forests act as the protective shield for them. The socio-cultural life of Saharia is greatly connected to the forests and forest products. The indigenous knowledge is inextricably linked with bio-diversity of the region, especially in the case of their medical system. Their knowledge of ethnomedicine (traditional medical system) is deep rooted in the ecology and culture. The flora and fauna has a significant place in the lives of the Saharia. Thus there exits a relationship between ecology, culture and health care practices among Saharia.

It can be said that every culture, irrespective of its simplicity and complexity, has its own beliefs and practices concerning diseases/illness and has evolved its own system of medicine in order to treat diseases its own way keeping the socio-cultural and environmental factors. It is, therefore, important to study in-depth the various ecological and socio-cultural practices affecting the health status of the Primitive Tribal Groups (PTGs).

Suggestions

In most of the tribal communities, there are number of folklores related to health. Certain practices are suggested to avoid illness or diseases, while some are prescribed to have better health. These should not be ignored as mere folk beliefs, but need careful attention. There are different folklores to avoid illness, during illness regarding food and so on. Documentation of such folklores available in different socio-cultural system may be very rewarding and may provide the model for appropriate health and sanitary practices in a given eco-system.

The sincere, missionary and concentrated effort of a medical anthropologist to learn and understand the health problems of tribals qualifies him as the only "outsider" able to give concrete and thorough recommendations to health providers working in tribal areas. The role of medical anthropologist with integrated-holistic approach is unique among those (health providers and bureaucrats) who deal with the tribals and it is hoped that the bureaucrats should appreciate and consider the suggestions made by him in respect of tribal welfare. The anthropologist has certain advantages over the administrators who are working in tribal areas. The anthropologist resides for a long period of time in the village to know and understand in an intimate way the people with whom he works. Administrators hardly ever find time to spend more than occasional moments among the people for whose welfare they are responsible. The researcher is of the opinion that the tribal development needs special and immediate attention of planners and administrators in general and anthropologists in particular for the development of a tribe. There is a need for co-ordination between administrators, anthropologists, Government officials, planners, political leaders, social workers, tribal leaders and NGO workers.

Great attention and importance need to be given to the protection and management of the environment. In the villages it is quite apparent that education is poor, other facilities like sanitation, drinking water, drainage, roads etc., are inadequate and the development of these facilities are not commensurate with the rural environment. Environmental education should be emphasized in the school curriculum. There is a dire need of painstaking programmes of health education among the Saharia to improve their health status. There are problems of living in difficult terrains, inadequate communication, transport facilities etc.

The Govt. should try to make Saharia conscious about better standard of life. Saharia should be made to realise that no programme can be successful until and unless they try to develop themselves. The Govt. should check the process of exploitation. The Govt. should take care of the funds and its utilization pattern in the programmes for the welfare of the Saharia. It should be observed whether the Saharia are aware about these programmes or not. There should be development of transportation especially in the interior villages. The Govt. should observe that whether the benefits reach the Saharia or not. The Govt. should also try to have a feedback about the effects of a programme. This feedback will indicate whether the programme is fully effective or not and if not, what type of changes are required therein or the current programme can be stopped and new programme should be started. Education is the most important vehicle that opens prospects for good health. The Govt. should improve the education and literacy status of the Saharia because education will create awareness about personal and community hygiene, about the health needs of the mother and the child, about nutrition and immunization. A large majority of water borne diseases can be eliminated by providing safe drinking water to the Saharia. They need preventive measures more than curative methods of health care.

Three major sources for treatment are used by the Saharia which include government health service centres, private practitioners and traditional healer or tribal medicine man. The Saharia combine both modern and traditional medicine according to their needs and choice. Given this situation, it is appropriate to promote both the systems, keeping in mind that these are not opposed to each other as far as the interests of the tribal people are concerned. Let the people choose what is best suited to them between the two. The Saharia don't give preference to any particular method of treatment. They have equal faith in all the treatments (spiritual, herbal and allopathic) and Government's emphasis should be more on the preventive medicine, than curative which is very expensive for the poor tribals.

Young boys and girls from the Saharia community should be given training in Public Health and Mid-wifery respectively. They may regularly visit the 'Saharana' and advise the people on health, sanitation and the necessity of availing modern medical attention in time. A serious and missionary spirit is required to improve the health status of the Saharia because they are illiterate and backward. They need guidance to make them aware of the merits and demerits of a system. For this all the concerned must apply themselves religiously drinking the nectar from the fountain of knowledge in whatever form it may come.

Health is a function not only of medical care but of overall integrated development of society — cultural, economic, educational, social and political. Good health and good society go together. This would be possible only when a number of supportive services such as nutrition, improvement in environment and in education go up to a very high level. In this task special care should be taken for provision of adequate health care for the Saharia. The major health problems can be avoided by ensuring safe drinking water, curative and preventive health care, health education etc. to Saharia. Despite efforts for raising the economic standard the Saharia fate remained more or less the same. The development effort did not take cognizance of urgent need of Saharia development. The National Health Policy was formulated and adopted in 1983. Since then there have been major changes in disease profile, healthcare infrastructure and healthcare seeking behaviour of the population. Several newer technologies for diagnosis and management of health problems have become available. These, in turn, have widened the gap between what is possible and what is feasible and affordable at the level of the individual and the country. Increasing expectations of the population, rising cost of diagnosis and treatment and diminishing resources have brought into fore the issue of how to meet the rising healthcare costs. The essential inter-linkages between health services delivery and health manpower development are still not fully understood and operationalised. Taking all these into consideration it is essential that the National Health Policy undergoes a re-appraisal and re-formulation so that it provides a reliable and relevant policy framework not only for improving healthcare, but also measuring and monitoring the healthcare delivery systems and health status of the population during the next two decades. It is neither wise nor expedient to interfere with their moral virtues, self-reliance and optimistic joy of their life, too hastily, unless they come forward with due incentives to integrate themselves of their own accord into higher values of modern life. In devising their policy the main criteria should be their acceptability and identification with people's way of life and not what we imagine to be good for them. Right for self-determination should be given priority and persuasion and participation should be the working tools and carefully planned policy based on scientific study should be the guiding principle in the cautious formulation of welfare plans with a view to accommodating tribal needs with regional and national interests. Hunger, disease, exploitation, ignorance, isolation are evils whose cure cannot be delayed, they must be treated rapidly and efficiently. People are responsible for their own development. Change is not something to be imposed from outside. People must participate in the process of their own development; i.e. they must be able to identify their own problems, utilise the available services for solving the same and imbibe appropriate knowledge and own quality of life. Change in the behaviours of people is possible only when it is community based i.e. it is all pervasive in the community.

Future Strategies

There is a need to reorganize, restructure and reallocate resources for Saharia community, based on their requirements. The design of the development programmes and the health needs of the Saharia has to be identified keeping in mind their ecology, socio-cultural practices and their knowledge of indigenous medicine. Mearly increasing the numbers of the health care facilities and inflow of technical manpower is not going to solve the health problems of Saharia. There is a need to train the paramedical staff posted in the tribal area to understand the specific health disorders occurring in the Saharia community and be able to treat them. The health education regarding immunization, maternal, reproductive and child health should be taken very seriously.

The health system need to be redesigned keeping in mind the concept of health and disease existing among the Saharia; the concerns e.g, equity, coverage, effectiveness, efficiency, impact; the objects, e.g. health centres, health programmes, and persons, e.g. providers, and consumers. There is a need to develop a "people- oriented" alternative health delivery system for the Saharia. Clearly such an alternative system cannot be floated merely on a new package of actions or with a new technology fix. It has to be built on a new vision of health care, with a new attitude of mind, coupled with a new value orientation in health delivery. To make this new alternate system a reality efforts are require to imbibe the concept of community and improve its spirits, and work should be done on various dynamics that are in operation at the moment. More importantly, the Saharia should be helped, because without helping and understanding them there is little chance of effecting any meaningful improvement in the health status of Saharia. Mere increase in the amount of medicine or, in medical person availability or, enhancement in the number of hospital beds, or increase/ improvement of any other such artifacts are unlikely to bring results for them. There is a need to integrate health actions with developmental welfare and educational activities. Health actions can really produce good results when they are coupled with developmental welfare and educational activities. Due to poor socio-economic status of Saharia and lack of very basic facilities, like, clean drinking water, environmental sanitation, etc. there is a need to provide them a parallelely four types of care (curative, preventive, promotive and rehabilitative activities in health action) and that too at the same time. There is a need for integrating curative with preventive, promotive, and rehabilitative activities in health actions. There is a need to integrate various systems of medicine existing among the Saharia (ethnomedical system, herbal medicine and allopathic system). A meaningful integration of the various systems could perhaps do a lot of good in improving physical, mental, social and emotional health of the Saharia. The community institutions like Saharia Panchayat and Local leaders should be involved in coping up with various demands of the Saharia health, health planning, service delivery and development. If the policy makers work on the above mentioned outlines then definitely there will be an improvement in the health status of the Saharia and this work will serve its basic purpose.

References

Ackernecht, E.H.: Problems of Primitive Medicine. *Bulletin of History of Medicine*, 11: 503-521 (1942).

Adhikary, A.K.: Hunter-Gatherers in India: A Preliminary Appraisal of Their Survival Strategy. ICSSR, Guwahati (1992).

Adolph, E.F.: *Physiology of Man in the Desert*. Inter-Science Publishers. New York (1947).

Ahuja D.R.: *Folklore of Rajasthan*. National Book Trust of India. New Delhi (1980).

Ali, A.: Health and Genetic Problems of Kutia Kondhs of Burlubaru Village Phulbani District, Orissa. *The Newsletter* (Government of India, Ministry of Home Affairs, Tribal Development Division, New Delhi) 1:103-114 (1980).

Ali, A.: Nutrition. In: *State of India's Health*. A. Mukopadhyay (Ed.).Voluntary Health Association of India (VHAI). New Delhi (1992).

Ali, A.: Indigenous Health Practices Among Tribals: Relationship with Prevalent Diseases. pp. 130-150 In: *Tribal Health in India*. S. Basu (Ed.). Manak Publication. New Delhi (1994).

Alland, A., Jr. Medical Anthropology and the study of biological and cultural adaptations, *American Anthropologist* 68(1): 40-51 (1966).

Alle, W.C., Emerson, A.E., Park, O., Park, T and Schmidt, K.P.: *Principles of Animal Ecology*. W.B.Saunderer Company. Philadelphia (1949).

Anderson, R.K., Calvo, Jose., Serrano,Gloria.,and Payne, G.C.: A Study of the Nutritional Status and Food Habits of Otomi

Indians in the Mezquital Valley of Mexico, *American Journal of Public Health,* 36:883-903 (1946).

Aiyappan, A.: *Report on the Socio-Economic conditions of the Aboriginal Tribes of the Province of Madras*, Government of Madras (1948).

Bailey, K.D.: *Methods of Social Research*. The Free Press. New York (1978).

Banerji,A.: Culture Sequence in Rajasthan. *Man in India*, 43: 225-232 (1963).

Banerji, B.G.: *Socialization: An Ethnic Study*. Inter-India Publications. New Delhi (1986).

Banerji,B.G and Bhatia,K.: *Tribal Demography of Gonds*. Gian Publishing House. Delhi (1988).

Banerji, D.: *Poverty, Class and Health Culture in India*. Prachi Prakashan. New Delhi (1982).

Banerji, D.: *Social Science and Health Service Development in India*. Lok Prakash. New Delhi (1986).

Banks, A.I.: *Man and His Environment*. University Press. Cambridge (1950).

Barrows, H.H.: Geography as Human Ecology. *Annals of the Association of American Geographers*. 13:1-14 (1923).

Barth, F.: Ecological Relationship of Ethnic Groups in Swat. North Pakistan, *American Anthropologist*, 58: 1088 (1956).

Basu, A.: Anthropological Approach to Tribal Health. pp. 131-142 In: *Tribal Demography and Development in North-East India*. A. Bose, T. Nangbri and N. Kumar (Eds.). B.R. Publishing Corporation. Delhi (1990).

Basu, K. and Mitra, K.: Health Practices and Problems of Slum Women of Delhi. *South Asian Anthropologist*, 19: 37-43 (1998).

Basu, S.K.: Genetics, Socio-Cultural and Health Care among Tribal Groups of Jagdalpur and Konta Tehsils of Bastar District, M.P. pp. 87-105 In: *Anthropology, Development and National Building*. A.K. Kalla and K.S. Singh (Eds.). Concept Publishing Company. New Delhi (1986).

Basu, S.K.: Determinants of Health Seeking Behaviour among the Tribal Population of Bastar District, Madhya Pradesh, India. *South Asian Anthropologist*, 11: 1-6 (1990).

Basu, S.K. Health and Culture Among the Underprivileged Groups in India. In: *State of India's Health*. A. Mukhopadhyay (Ed.). Voluntary Health Association of India (VHAI). New Delhi (1992).

Basu, S.K. Health Problems and Health Care of the Tribal Population of India. pp. 137-146 In: *Tribal Economy and Wasteland Development*. N. Mahanti (Ed.). Inter-India Publications, New Delhi (1994).

Basu, S.K. Need for Action Research for Health Development in Tribal Communities of India. *South Asian Anthropologist*, 17: 73-80 (1996).

Basu, S.K.: *Genetic Disorders and Health Care*. Shri Kala Prakashan. Delhi (1994).

Basu, S.K.: (Ed.).*Tribal Health in India*. Manak Publications. Delhi (1994).

Basu, S.K. and Kshatriya, G. K.: Fertility and Mortality in Tribal Population of Bastar District,M.P. *Biology and Society*, 6: 110-112 (1989).

Basu, S.: *Health Need Assessment of Tribal Populations of District Midnapur, West Bengal*. Ph.D.Thesis. Unpublished. University of Delhi, Delhi (1998).

Bates, M.: Human Ecology. pp. 700-713 In: *Anthropology Today*. A.L. Kroeber (Ed.).University of Chicago Press, Chicago. (1953).

Bates, M.: The Ecology of Health. pp. 56-77 In: *Medicine and Anthropology*. I. Galdston (Ed.). International University Press. New York (1959).

Beattie, J.: *Other Cultures*. Cohen and West Limited. London (1965).

Behura, N.K.: Anthropology and Development. In: *Tribes and Tribal Life*. Singh, A.K (Ed.) Vol. II. Sarup and Sons. New Delhi (1996).

Bernam, P.C. Kendall and K.Bhattacharya.: The Household Production of Health: Integrating Social Science Perspective on Micro Level Health Determinants. *Social Science and Medicine,* 38:205-215.

Bews, J.W.: *Human Ecology*. H. Milford, London (1935).

Bhasin, M.K. *Cold Desert: Ladakh, Ecology and Development*. Kamla-Raj Enterprises. Delhi (1992).

Bhasin, M.K. and Bhasin, V.: *Sikkim Himalayas: Ecology and Resource Development*. Kamla-Raj Enterprises. Delhi (1995).

Bhasin, M.K. and Bhasin, V.: *People of Rajasthan:Ecology,Culture and Society*. Kamla-Raj Enterprises. Delhi (2000).

Bhasin, V.: *Habitat,Habitation and Health in the Himalayas: A Comparative Study of the People of Sikkim and the Gaddis of Himachal Pradesh*. Kamla-Raj Enterprises. Delhi (1990).

Bhasin, V. and Bhasin, M.K.: Ecology, Population Dynamics among Transhumant of Himalayas. A case study of Gaddis of Himachal Pradesh. *Journal of Human Ecology*, 4:171 (1993).

Bhasin, V.: *People, Health and Disease: The Indian Scenario*. Kamla-Raj Enterprises. Delhi (1994).

Bhasin, V.: Religious and Cultural Perspectives of a Sacred Site – Sitabari in Rajasthan. *Journal of Human Ecology*, 10: 329-340 (1999).

Bhasin, V.: *Tribals of Ladakh: Ecology, Human Settlements and Health*. Kamla-Raj Enterprises. Delhi (1999).

Bhasin, V. Profile of Rajasthan.pp.3-20 In: *People of Rajasthan: Ecology, Culture and Society*. M.K. Bhasin and V. Bhasin (Ed.) Kamla-Raj Enterprises. Delhi (2000).

Bhattacharya, P.: Folk Medicinal Claims for the treatment of Veneral Diseases from Patalkot Valley of Chhindwara District (M.P), *South Asian Anthropologist*, 16: 19-21 (1995).

Bhattacharya, S. and Sengupta, S.K.: The Concept of Disease and Its Treatment among the Birhor of Purulia. In: *Tribal Health: Socio-Cultural Dimensions*. B. Chaudhuri (Ed.). Inter-India Publications. New Delhi (1986).

Bhil, R.S.: A Preliminary Appraisal of the Saharia Totemism. *Vanyajati*, Tribal Research Institute, Udaipur, (1973).

Bhowmick, P.K.: *The Lodhas of West Bengal: A Socio-Economic Study*. Punthi Pushtak. Calcutta (1963).

Bidney, D.: *Theoretical Anthropology*. Columbia University Press, New York (1953).

Binford, L.: Post Pleistocene Adaptations. pp. 313-341. In: *New Perspectives in Archaelogy*. Lewis R. Binford and Sally Binford (Eds.). Academic Press. New York (1968).

Blumenhagen, D.W.: Hypertension: A Folk Illness with a Medical Name. *Culture, Medicine and Psychiatry, 4: 197-227* (1980).

Blumenhagen, D.W.: On the nature of explanatory models. *Culture, Medicine and Psychiatry*, 5:331-340 (1981).

Bose, A.: *Demographic Diversity of India. State and District level Data. A reference Book*. B.R. Publishing Corporation. Delhi (1991).

Bose, A.B. and Saxena, P.C.: The Diffusion of Innovations in a Village in Western Rajasthan, *Eastern Anthropologist*, 18: 138-145 (1963.).

Bose, N.K.: Change in Tribal Cultures Before and After Independence. *Man in India*, 44 (1964).

Brown, A.R.: *Structure and Function in Primitive Society*. Cohen and West Limited. London (1959).

Burgess, G.R. Approaches to Field Research. In: Field Research: A Sourcebook of Field Manual. George.G.Burgess (Ed.). Academic Division Unwin Hyman Limited. London (1982).

Burnham, P.C. and R.F. Ellen.: *Social and Ecological System*. London (1979).

Cage, T.J.: Theories Differ on Use of Focus Groups. *Advertising Age*, 19: 20-22 (1980).

Calder, B.J.: Focus Groups and the Nature of Qualitative Marketing Research. *Journal of Marketing Research*, 14: 353-364 (1977).

Carstairs, G.M.; Medicine and Faith in Rural Rajasthan. In: *Health, Culture and Community*, Benjamin D.Paul (Ed.). Russel Sage Foundation. New York (1955).

Catford, J. C.: Positive Health Indicators Towards a New Information Base for Health Promotion. *Community Medicine*, 5: 123-132 (1983).

Caudill, W.: Applied anthropology in Medicine. In: *Anthropology Today*. A.L. Kroeber (Ed.).University of Chicago Press. Chicago (1953).

Census of India, Rajasthan, Vol XIV, Part VI Government of India, (1961).

Census of India, Scheduled Castes and Scheduled Tribes, Registrar General and Census Commissioner of India, (1991).

Census of India, Rajasthan, District Census Handbook, Kota District, Kota (1981).

Census of India, Primary Abstract Series 1, Scheduled Tribe Population, Office of Registrar General and Census Commissioner, India (1991).

Census of India, Rajasthan, District Census Handbook, Kota District, Kota (1991).

Census of India, Rajasthan, Series 21, Directorate of Census Operations, Rajasthan, Jaipur (1991).

Chakravarty, M and Singhrol, C.S.: Socio-Cultural and Environmental Aspects of Health and Nutrition. pp. 9-12. In: *Anthropology: Some Socio-Demographic and Nutritional Aspects of Health*. A.N. Sharma (Ed.). Northern Book Centre. New Delhi (1989).

Chatterjee, K.K.: Health Status of Tribal Women, *Social Change*, 23(4) 53-55. (1993).

Chaudhuri B.: Concept and Treatment of Disease among the Tribals of India.

ICSSR Review Study (1985).

Chaudhuri B.: Social and Cultural Aspects of Health. *Journal of Social and Economic Studies*, 39: 379-388. (1986).

Chaudhuri B.: Social Science, Health and Culture: The Tribal Situation. *Social Action*, 39: 243-253. (1986).

Chaudhuri, B.: Science in Society: The Case of Tribal Medicine. *Seminar. The Philosophy of Science,* Indian Institute of Advanced Study, Shimla (1990).

Chaudhuri, B.:(Ed.). *Tribal Health: Socio-Cultural Dimensions.* Inter-India Publications. New Delhi (1990).

Chaudhuri, B.:(Ed.). *Cultural and Environmental Dimensions of Tribal Health.* Inter-India Publications. New Delhi (1990).

Chaudhri B.: Tribal Health and Medicine: Socio-Cultural and Environmental Dimensions.pp124-136 In: *Tribal Economy and Wasteland Development. N.*Mahanti (Ed.). Inter-India Publications. New Delhi. (1994).

Chaudhuri, N.D.: *Tribal Development in Rajasthan: Past Efforts and New Challenges*, MLV Tribal Research and Training Institute,Udaipur (1986).

Chauhan, B.R.: Rajasthan-Ke Ek Gaon Mein Manav-Ka-Divya Shaktiyon Se Sambandh, *Samaj*, Khandelwal Press. Banaras (1956).

Clarke, R. and Hindley,G.: *The Challenge of the Primitives*. McGraw Hill. New York (1975).

Colson, A.C.and Selby, K.E.: Medical Anthropology. *Annual Review of Anthropology*, 3:245-262 (1974).

Comaroff, J.: Medicine Symbol and Ideology. In: *The Problems of Medical Knowledge*. P. Wright and A. Treakher (Ed.). Edinburg University Press. Edinburg (1982).

Crook, W.: *The Tribes and Castes of North-Western India*. Cosmo Publications. Delhi (1975).

Corwin, E.H.L.: *Ecology of Health*. (New York Academy of Medicine Centennial, Institute of Public Health,1947) Common Wealth Fund. New York (1949).

Dash, J.: The concepts and treatment of Disease in Tribal Orissa In: *TribalHealth: Socio-Cultural Dimensions*. B.Chaudhuri (Ed.). Inter India Publications, New Delhi (1986).

Dave, P.C.: *The Grasias*. Bhartiya Adim Jati Sewak Sangh. Kingsway. Delhi (1960).

Deogaonkar, S.G.: *Tribal Development Plans – Implementation and Evaluation*. Concept Publishing House. New Delhi (1992).

Davis, K.: *Human Society*. MacMillan. New York (1948).

Devarapalli, P.S.: Maternal Care and Obstetric Practices among the Konda Gora Tribe. *Journal of Family Welfare,* 39: 44 (1996).

De Young, J.E.: *Village Life in Modern Thailand*. University of California Press. Berkeley (1995).

Dey, I.: *Qualitative Data Analysis: A User Friendly Guide for Social Scientists*. Routledge. New York (1993).

Debhar, U.N.: *Report of Scheduled Tribe Commission,* Government of India, New Delhi (1961).

Doshi, S.L. and Vyas, M.N.: *Tribal Rajasthan: Sunshine on the Aravali*. Himanshu Publication. Udaipur (1992).

Dube, S.C.: *The Kamar*. Universal Publishers Limited. Lucknow (1951).

Dube, S.C.: *Indian Village*. Routledge and Kegan Paul Limited. London (1955).

Dube, S.C.: *Tribal Heritage of India: Ethnicity, Identity and Interaction*. Vikas Publications. New Delhi (1977).

Economic Review, Directorate of Economics and Statistics, Rajasthan, Jaipur (1997-98).

Elton, C.: *Animal Ecology*. Macmillan Company. New York (1927).

Elwin, V.: *The Baiga*. John Murray. London (1939).

Elwin, V.: *The Aboriginals*. Oxford University Press. Bombay (1943).

Elwin, V.: *Bondo Highlanders*. Oxford University Press. Bombay (1950).

Elwin, V.: *The Religion of an Indian Tribe*. Oxford University Press. London (1955).

Elwin, V.: *Philosophy for NEFA*. Shillong (1959).

Evans-Pritchard, E.E.: *The Nuer: A Description of the Modes of Livelihood and Political Institutions of a Nilotic People*. Oxford University Press. London (1932).

Evans- Pritchard, E.E.: Witchcraft, Oracles and Magic Among Azande. Oxford University Press. London (1937).

Evans- Pritchard, E.E.: Some Reminiscences and Reflections on Fieldwork. *Journal of the Anthropological Society of Oxford, 4: 1-12* (1973).

Fabrega Jr., Horacio: Medical Anthropology. pp. 167-229. In: *Biennal Review of Anthropology*. B,J. Siegal (Ed.).Stanford University Press (1972).

Fazl, A.: *Akbar Nama*, Kitab Publishing House. Delhi (1917).

Feyisetan, K.: *Focus Groups, paper presented at Quantitative Methods Workshop, Ifakara, Tanzania,* 11-29 April, 2000 (2000).

Firth, R.: *Elements of Social Organisation*. Watts. London (1951).

Firth, R.: (Ed.) *Man and Culture*. Routledge and Kegan Paul. London (1957).

Firth, R.: *Primitive Polynesian Economy*. Routledge and Kegan Paul. London (1965).

Firth, R.: Engagement and Detachment: Reflections on Applying Social Anthropology to Social Affairs. *Human Organisation*,40 (1981).

Forde, D.: Spirits, Witches and Sorcers. *Journal of Royal Anthropological Institute of Great Britain and Ireland*, 88 (1958).

Fortes, M.: Introduction. In: *Social Anthropology and Medicine*. Loudon (Ed.). Academic Press. London (1976).

Foster, G.M. and Anderson, G.A.: *Medical Anthropology*. John Wiley and Sons. New York, Bribane and Toronto (1978).

Frake, C.: The Diagnosis of Disease Among The Subanum of Mindanas. *American Anthropologist*, 63: 113-132 (1961).

Friedson, E.I.: The Sociology of Medicine: A Trend Report and Bibliography. *Current Sociology*.10-11 (1961-62).

Fuchs, S.: Magic-Healing Techniques Among the Balahis in Central India. pp. 121-138 In: *Magic,Faith and Healing: Studies in Primitive Psychiatry Today*. Arikiev (Ed.). Free Press. Glencoe (1964).

Fuchs, S.: *The Aboriginal Tribes of India*. Macmillan, Delhi (1992).

Geertz, C.: *Agricultural Involution: The Process of Ecological Change in Indonesia*. University of California Press, Berkeley (1963).

Geertz, C.: *The Interpretation of Cultures*. Basic Books. New York (1973).

Gehlot, J.S.: *The History of Rajasthan*. Vol. I Rajasthan Sahitya Mandir (1981).

Ghatak, N.K.: Saharia. In: People of India (Rajasthan). K.S. Singh, D.K. Samanta, S.K. Mandal and N.N. Vyas (Ed.). Anthropological Survey of India (1994).

Ghatak, N.K. and Nanda, D.K.: The Saharia of Rajasthan: Impact of Rural Development Programme. *Vanyajati,*Tribal Research Institute (TRI), Udaipur, 22-26 (1986).

Ghoushmalik, S.K. and Karmohapatra,R.K.: *Health Situation of the Juangs of Orissa*. South Asian Anthropologist,16: 79-83 (1995).

Ghurye, G.S.: *The Aborigine's-so-called- and their Future*. Gokhale Institute of Politics and Economics. Poona (1943).

Ghurye, G.S.: *The Scheduled Tribes*. Popular Book Depot. Bombay (1959).

Good, B.J.: The Heart of what's the matter: the semantics of illness in Iran. *Culture,Medicine and Psychiatry,* 1:25-58 (1977).

Gould, Harold A.: The Implications of Technological Change for Folk and Scientific Medicine. *American Anthrropologist*, 59:507-516 (1954).

Government of Rajasthan.: *Report on Saharia Development Plan*. Tribal Area Development Department (1979-80).

Griegson, W.G.: *The Maria Gonds of Bastar*. Oxford University Press. London (1938).

Gross, E.: Social Science Techniques: A Problem of Power and Responsibility. *Scientific Monthly*, 83:242 (1956).

Gullick, J.: Village and City Fieldwork in Lebanon.pp.89-118 In: Marginal Natives At Work: Anthropologists in the Field. M. Freilich (Ed.) Wiley, New York (1977).

Gupta, R.: *Sanwara: A Village Survey*. Census of India (1961).

Gupta, R.: Selection Intensity in Sherpas. *Current Anthropology*, 21:136-137 (1980).

Gupta, S.P.: Tribal Concepts on Health, Disease and Remedy. pp. 161-166 In: *Tribal Health: Socio-Cultural Dimensions*. B. Chaudhuri (Ed.).Inter-India Publications, New Delhi (1986).

Hackett, OL.W.: *Malaria in Europe: An Ecological Study*. H. Milford. London (1937).

Hahn, Robert A. and Kleinman,Arthur.: Biomedical practice and anthropological theory: Frameworks and directions. Annual Review of Anthropology, 12:305-333 (1983).

Haimendorf, C.V.F.: *The Chenchus: Jungle Folk of Deccan*. MacMillan and Company. London (1943).

Haimendorf, C.V.F.: *The Reddis of Bison Hills: A Study in Acculturation*. MacMillan and Company. London (1945).

Haimendorf, C.V.F.: *The Raj Gonds of Adilabad. MacMillan and Company*. London (1948).

Haque,M.: Height, Weight and Nutrition among the Six tribes of India. pp. 192-206. In: *Cultural and Environmental*

Dimensions of Health. B. Chaudhuri (Ed.). Inter-India Publications. Delhi (1986).

Hardesty, D. L.: *Ecological Anthropology*. John Wiley. New York (1977).

Harper, E.B.: Shamanism in South India. pp. 344-353 In: *Sociology of Medical Institution*. R. Scott and E.H. Volkart (Eds.). John Wiley and Sons. New York (1966).

Hasan,K.A.: *Cultural Frontiers of Health in Village India*. Manaktalas. Bombay (1967).

Haug, M.R., M.L. Wyke and K.H. Namazi.: Self Care among Older Adults. *Social Science and Medicine*, 29:171-183 (1989).

Hawley, A.H.; *Human Ecology: A Theory of Community Structure*. Ronald Press Company, New York (1950).

Herskovits, M.J. *Man and His Works*. Knopf. New York (1948).

Herskovits, M.J. *Cultural Anthropology*. Oxford and IBH Publishing Company. New Delhi (1951).

Hiramani, A.B.: *Cultural Correlates of Tribal Health*. B.R. Publishing Corporation. New Delhi (1997).

Hughes, Charles, C.: Ethnomedicine. pp. 87-92. In: *International Encyclopaedia of Social Sciences*. David Sills (Ed.). Crowell Collier and MacMillan (1968).

ICMR Bulletin: Highlights of Tribal Health Research under ICMR. *ICMR Bulletein Vol. 26* March-April (1996).

Jaggi, O.P.: *Folk Medicine*. Atma Ram. New Delhi (1973).

Jain, K.C.: *Ancient Cities and Town in Rajasthan*. Delhi (1972).

Jain, S.K.: *Medicinal Plants*. National Book Trust. New Delhi (1966).

Jaspan, M.A.: Health and ILLth in Highland Sumatra. In: *Social Anthropology and Medicine*. J.B. Loudon (Ed.). Academic Press. London (1976).

Jay, E.: The Anthropologist and Tribal Welfare: Hill Maria – A Case Study. *Journal of Social Research*, 2: 82-89 (1929).

Jay, E.: *A Tribal Village of Middle India*. Anthropological Survey of India, Calcuttta (1970).

Joshi, P., Khandelwal, S., and Shrivastava,Y.: Drops of Plant Conservatioin in Rajasthan: Sacred Grooves. *Eco-*

Development, Habitat and Wildlife Conservation in Rajasthan and Adjoining Areas. Hadoti Naturalist's Society, Kota. 40-43 (1994).

Joshi, S.G.: *Medicinal Plants.* Oxford and IBH Company Private Limited. New Delhi (2000).

Joshi, V.: *Submerging Villages.* Ajanta. New Delhi (1987).

Joshi, V.: *Tribal Situation in India: Issues in Deveopment.* Rawat Publications. Udaipur (1998).

Kahn, R.L. and Cannell, C.F. *The Dynamics of Interviewing: Theory Technique and Cases.* Wiley. New York (1957).

Kapoor, A.K. and Prasad, R.R.: A Note on Beliefs and Practices Pertaining to Child-Birth Among the Bhotias. *Vanyajati,* 4:35-37 (1982).

Kapoor. A.K.: Environment and Settlement Pattern Among the Bhotias of Central Himalayas. In: *Wild Life Under Threat* (Ed.). S.K. Chadha. Vinod Publishers. Jammu (1991).

Kapoor A.K.: Ecology and Tribals in Central Himalayas. In: *Ecology and Man in the Himalayas (Eds.).* A.K. Kapoor and S. Kapoor. M.D. Publication. New Delhi (1994).

Kapoor A.K.: Bio-Social Study of Saharia: A Primitive tribe of Rajasthan. *Project Report submitted to University of Delhi,* Delhi (1996).

Kapoor. A.K and Singh, D.: *Rural Development Through NGOS.* Rawat.Publications, Jaipur (1997).

Kapoor, A.K.: Medical Ecology and Settlement Pattern in a Himalayan Tribe. *Indian J. Community Medicine (communicated)* (1997).

Kapoor, A.K.: The Anthropologist as a Bio-Social Hero: Health Medicine and Magic in a Scheduled Tribe of Himalayas Paper presented at *National Seminar on Health and Hygiene among the Tribal Communities in India. Dec 15th-17th, 1998 organised by Department of Anthropology,Andhra University,Visakhapatnam* (1998).

Kapoor, A.K.: Ecology, Disease Pattern and Healing Aspect in Himalayan Communities. *Man in India (communicated)* (1999).

Kapoor, A.K.: Environment,Health and Development in a Primitive of Rajasthan. pp. 103-133 In: *Environment,*

Health and Development. P. Dash Sharma (Ed.). S.C. Roy Institute of Anthropological Studies, Ranchi (2000).

Kapoor, A.K.: Social Functions and Ceremonies in Koya Tribal of District Koraput Orissa In: *Ethnographic Atlas of Tribals*. Prakash Chandra Mehta (Ed.). Shiva Publishers, Jaipur (2000).

Karna, M.N.: Etiology of Disease: A Sociological Study in the Rural Context *Man in India,* 56 (1976).

Kashyap,A. and Kumar, R. 1998. Tribal Rajasthan: A Scenario. pp. 41-55. In: *Tribal Situation in India: Issues and Development*. V.Joshi (Ed.) Rawat Publication. New Delhi.

Khan, G.: Tribal Health: Impact of Rural Development Programme. pp. 311-321 In: *Tribal Health: Socio-Cultura Dimensions*. B.Chaudhuri (Ed.).Inter-India Publications, New Delhi (1986).

Khan,G,A.: The Chenchus. In: *Essays in Anthropology*. J.P.Mills (Ed.). Maxwell and Company. Lucknow (1942).

Khare,R.S.: A Plea for Cultural Dimension in Medicine. *Eastern Anthropologist*,12 (1957).

Khera. P.D.: Ecology and Health. In: *Tribal Health in India*. S. Basu (Ed.) Manak Publications. New Delhi (1994).

Kleinman, A.: Concepts and Models for Comparison of Medical System. *Social Science and Medicine Bulletin*, 85-93 (1978).

Kleinman, A.: Clinical Relevance of Anthropological and Cross-Cultural Research: Concepts and Strategies. American Journal of Psychiatry, 135: 427-431 (1978).

Kleinman, A.: Patients and Healers in Context of Culture: An Exploration of Border Land Between Anthropology. *Medicine and Psychiatry*, University California Press. California (1980).

Kluckhohn, C.: *Culture and Behaviour: Collected Essays*. Free Press. New York. (1962).

Kluckhohn, F.: The Participant Observer Technique in Small Societies. *American Journal of Sociology*. 46:331-343 (1991).

Kroeber, A.L.: *Cultural and Natural Areas of Native North America*. University of California Press. Berkeley (1939).

Kshatriya, G.K.: Ecology and Health With Special Reference To Indian Tribes. pp. 229-245. In: *Man-Environment*

Relationship. M.K.Bhasin and Veena Bhasin (Ed.). Kamla-Raj Enterprises. Delhi (2000).

Kshatriya, G.K., Singh, P. and Basu, S.: Fertility and Mortality in Bison Horn Madias of Dantewara tehsil of Bastar District of MP. *Journal of Human Ecology,* 4:93 (1993).

Kshatriya, G.K., Singh, P. and Basu, S.: Anthropo-demographic Features and Health Care Practices among the Jaunsaris of Jaunsar-Bawar, Dehradun, UP. *Journal of Human Ecology*, 8:347 (1997).

Kumari, P.: Herbal Medicine among the Mundas of Chottanagpur. *Tribal Research Bulletin,* 12: 12-20 (1990).

Lakshminarayana.: Health Conditions in Tribal Areas of Maharashtra State. *Tribal Research Bulletin,* 12: 1-4 (1990).

Landy, D. (Ed.). *Culture, Disease and Healing: Studies in Medical Anthropology*. MacMillan Publishing Company. New York (1977).

Leavell, H.R. and Clark, E.G.: *Preventive Medicine for the Doctor in his Community*. McGraw Hill Book Company. New York (1965).

Leiben, R.W.: Medical Anthropology. In: *Handbook of Social and Cultural Anthropology*. J.J.Honigman (Ed.). McNally College Publishing Company. Chicago (1973).

Leslie, C.: *Asian Medical Sytems*, University of California Press, Berkeley (1976).

Lewis, I.M.: Spirit Possession and Deprivation Cults. *Man, 3* (1966).

Lewis, O.: Controls and Experiments in Field Work .In: *Anthropology Today*. A.L.Kroeber (Ed.).University of Chicago Press, Chicago (1953).

Lewis, O.: *Village Life in Northern India*. Random House. New York (1958).

Lieban, R.W.: Medical Anthropology. In: *Handbook of Social and Cultural Anthropology*. J.J. Honigman (Ed.). Mcnelly College Publishing Company (1973).

Lieban, R.W.: The Field of Medical Anthropology. In: *Culture, Disease and Healing* . David Landy (Ed.). MacMillan. New York (1977).

Lock, M.: Models and Practice in Medicine Menopause as Syndrome of Life Transition. Culture. *Medicine and Psychiatry*, 6: 261-280 (1982).

Lowie, R H.: *Primitive Society.* Routledge and Kegan Paul Limited. London (1921).

Madge, J.: *Tools of Social Science.* Longmans Green and Company Limited. London. (1953).

Mahanta, K.C (Ed.).: *People of the Himalayas: Ecology, Culture, Development and Change*, Kamla-Raj Enterprises. Delhi (1997).

Mahapatra, L.K. and J. Das.: Nutritional Eco-systems of Orissa Tribal. In: *Cultural and Environmental Dimensions of Health.* B.Chaudhuri. (Ed.). Inter-India Publications. New Delhi (1990).

Mahanti, N and Singh, B.: (Eds.). *Tribal Health in India.* Inter-India Publications. New Delhi (1995).

Mahapatra, S.: Cultural Patterns in Health Care: A View from Santal World.pp. 249-258. In: Cultural and Environmental Dimensions of Health. (Ed.). B. Chardhuri. Inter-India Publications. New Delhi (1990).

Maheshwari, H.: *Rajasthan Bhasha Aur Sahitya.* Calcutta (1960).

Majumdar, D.N.: The Economic Background of Social Institutions in Primitive Societiey. *Man in India*,10: 150-166 (1930).

Majumdar, D.N.: Disease, Death and Divination in Certain Primitive Societies in India. *Man in India*, 13: 115-149 (1933).

Majumdar, D.N.: *A Tribe in Transition: A Study in Cultural Pattern.* Longmans Green and Company. London (1937).

Majumdar, D.N.: *Fortunes of Primitive Tribes.* Lucknow University Press. Lucknow (1944).

Majumdar, D.N.: *The Matrix of Indian Culture.* Universal Publishers Limited. Lucknow (1947).

Majumdar, D.N.: *Races and Culture of India.* Asia Publishing House. Bombay (1966).

Majumdar, D.N and Madan, T.N.: *An Introduction to Social Anthropology.* Asia Publishing House. Bombay (1956).

Malinowski, B.: *Culture. Encyclopaedia of the Social Science.*Vol IV, New York (1931).

Malinowski, B.: *Argonauts of the Western Pacific.*George Routledge. London (1932).

Malinowski, B.: *A Scientific Theory of Culture and Other Essays.* University of North California Press. Carolina (1944).

Mann, K.: *Tribal Women in a Challenging Society.* Mittal Publications. New Delhi (1987).

Mann, R.S.: *Social Structure, Social Change and Future Trends.* Rawat Publications. Jaipur (1979).

Mann, R.S.: Experimentation and Handicaps in Tribal Development. In: Indian Tribes in Transition. N.N. Vyas and R.S.Mann (Eds.). Rawat Publications. Jaipur (1980).

Mann, R.S. and K.Mann.: Religious Attributes and Cure in Tribal India. pp. 171-184 In: Tribal Health: Socio-Cultural Dimensions. B. Chaudhuri (Ed.). Inter-India Publications. New Delhi (1986).

Mann, R.S. and K.Mann.: *TribalCulturesandChange.* Mittal Publications, New Delhi (1993).

Mann, R.S.: *Culture and Integration of Indian Tribes.* M.D.Publication. New Delhi (1993).

Mann, R.S.: *Aspects of Indian Social Anthropology.* Concept Publishing Company New Delhi (1996).

Mann, R.S.: Economic Systems among Western Indian Tribes. In: *Economies of the Tribes and their Transformation.* K.S. Singh (Ed.). Concept Publishing Company. New Delhi (1982).

Marriott, M.: Western medicine in a Village of Northern India. pp. 239-269. In: *Health, Culture and Community.* B.D. Paul (Ed.), Russel Sage Foundation. New York (1955).

Marshall, C and Rossman, G.B.: *Designing Qualitative Research.* Sage Publication. London (1989).

Mathur, P.R.G.: Anthropology of Tribal Medicine: Disease and Curing Techniques among Tribes of North Wynad (Kerala). *Man in India,* 12 (1982).

Mathur, U.B.: *Ethnographic Atlas of Rajasthan.* Government Press. Jaipur (1969).

Matthews, C.M.E.: *Health and Culture in a South Indian Village*. Sterling. New Delhi (1979).

McDonal, C.: Focus Groups and their Role in the Marketing Research Process, *Akron Business and Economic Review*, 10: 14-19 (1979).

Mechanic, D. *Medical Sociology*. 2nd edition: A Comprehensive text, The Free Press. A Division of Macmillan Publishing Co., New York (1978).

Nickter, M.: The Lay Person's Perception of Medicine as Perspective into Utilization of Multiple Therapy System in the Indian Context. *Social Science and Medicine*. (1980).

Merring, V. and Kasdam,L.: *Anthropology and the Behavioural and Health Sciences*. University of Pittsburgh Press (1970).

Miri, M.:(Ed.). *Continuity and Change in Tribal Society*. Indian Institute of Advanced Studies. Shimla (1993).

Misra, P.: *Education in a Village Community: An Anthropological Study of Kaya Village in Udaipur District of Rajasthan*. M.Sc.Dissertation. Unpublished. University of Delhi, Delhi (1998).

Misra, P. and Kapoor, A.K.: Emerging Roles of Medical Anthropology for Health Providers who are Engaged in Tribal Development: A Study Among Saharia of Rajasthan. In: *The Science of Man in the Service of Man*. M.K. Bhasin and S.L.Malik (Eds.) Department of Anthropology. University of Delhi. Delhi (2001).

Misra, P. and Kapoor, A.K.: Role of Medical Anthropology in Tribal Development: Future Perspective. In: *Tribal Welfare and Development: Emerging Roles of Anthropological Explorations*. A.N.Sharama (Ed.) Sarup Publications. Delhi (2002a).

Misra, P. and Kapoor, A.K. Environment, Disease Pattern and Health Status in a Primitive Tribe of Rajasthan. In: *Eco-Degradation, Bio-Diversity and Population Health*. B.N. Pandey (Ed.). Daya Publishing House. Delhi (2002b).

Misra P. and Kapoor, A.K.: Ecology, Economy and Culture: An Anthropological Profile of Meena, a Scheduled Tribe of Rajasthan. *Social Change*. Council for Social Development, New Delhi (2002c).

Misra, P. and Kapoor, A.K.: Educating Tribal Girls. *Social Welfare. CSWB.* New Delhi (2002d).

Misra, P. and Kapoor, A.K.: Socio-Cultural and Environmental Factors of Health among Saharia: A Primitive Tribe of Rajasthan. In: *Contemporary Society: Tribal Studies* (Eds.) D.K.Behera and George Pfeffer (In Press).

Misra, P. and Kapoor, A.K.: Strategies for Health and Development of a Primitive Tribe: A Study in Rajasthan. In: B.S. Bisht and P.K. Pant (Eds). *Development of SC, ST and OBC*(In Press).

Misra, P. and Kapoor, A.K.: An Ethnographic Profile of Saharia: A Primitive Tribe of Rajasthan. In:*Ethnographic Atlas of Tribals,* P. C. Mehta (Ed.). Shiva Publishers. Jaipur (In Press).

Misra, P. and Kapoor, A.K.: Health and Development Policies for a Primitive Tribe: A study in Rajasthan. NIRD, Hyderabad (In Press)

Mitchell, D.G.A.: *New Dictionary of Sociology.* Routledge and Kegan Paul. London (1972).

Mitra, K. and Basu, K.: Health Problems and Practices of Slum Women of Delhi. *South Asian Anthropologist,* 19: 37-43 (1998).

Mohanty, D.: *Concern Areas in Tribal Women's Health in India: Tribal Health in India.* (Ed.). B.Singh and N. Mahanti, Inter-India Publications. New Delhi (1995).

Montagu, A.: *Anthropology and Human Nature.* Mac-Graw Hill Company. New York (1963).

Morgan, L.H.: *Ancient Society.* World Publishing Company. New York (1877).

Murthy, G.B.: The Soligas of B.R.Hills: A Demographic Study. *The Journal of Family Welfare.* 34: 54-58 (1987).

Nag, D.S.: *Tribal Economy: An Economic Study of Baiga.* Bhartiya Adim Jati Sewak Sangh, New Delhi (1958).

Naik, T,B.: (Ed.) The Abhuj Marias. *Tribal Research Institute.* Chhindwara (1963).

Naik, T,B.: Saharia: A Primitive Group of Our Neighbouring State. *Adivasi Gujarat,* 3:147 (1980).

Naik, T,B.: The Saharia. *Tribal Research and Training Institute, Gujarat Vidyapith*. Ahemadabad (1984).

Nath, Y.V.S.: *Bhils of Ratanmal. Maharaja Shivajirao*. University of Baroda, Baroda (1960).

National Family Health Survey (NHFS-2).Rajasthan. Internation Institute for Population Sciences. Mumbai (1998-99).

National Health Policy. Government of India. Ministry of Health and Family Welfare, New Delhi (1983).

National Institute of Communicable Diseases (NICD). Health Profile With Special Reference to Malaria and Yaws in Tribal District of Bastar MP. Annual Report (1993).

National Institute of Nutrition (NIN).ICMR, Hyderabad, Annual Report (1971).

National Population Policy. *Yojana*, 44: 57-70 (2000).

Nehru, J.: The Tribal Folk. *Adivasi*, Government of India, Ministry of Information and Broadcasting Publication Division, New Delhi (1955).

Nichter, M.: The Lay Person's Perception of Medicine as Perspective into Utilisation of Multiple Therapy System in the Indian Context. *Social Science and Medicine*, 4: 225-233 (1980).

Opler, M.E. The Cultural Definition of Illness in Village India. *Human Organization*, 22: 32-35 (1963).

Pacholi, S.: *Medical Geography of Malaria in Madhya Pradesh*. Northern Book Centre. New Delhi (1993).

Padhy, K.C.: *The Challenges of Tribal Development*. Sarup and Sons. Delhi (2000).

Pandey, G.D. and Tiwary, R.S.: Socio-Cultural Characteristics and Health Seeking Behaviour of the Hill Korwas. *Journal of Family Welfare*, 39:44 (1993).

Pandey, G.D., Roy and Tiwary, R.S.: Socio-Cultural Characteristics and Health Seeking Behaviour of the Khairwar of Madhya Pradesh. *Journal of Family Welfare*, 42: 49 (1996).

Parik, M.: Saharia. *Itwari Patrika of 28th December*,1975, Jaipur (1975).

Park, J.E.: *Text Book of Preventive and Social Medicine*. Banarsidas Bhanot. Napier Town, Jabalpur (1972).

Park, K.G.: Divination and Social Contexts. *The Journal of Royal Anthropological Institute of Great Britain and Ireland*. 90 (1963).

Parsuraman, S,and Rajan, S.: On the Estimation of Vital Rates Among the Scheduled Tribes in Western India. In: *Demography of Tribal Development*. A.Bose, U.P.Sinha and R.P. Tyagi (Eds.).(1990).

Patel, M. L.: *Tribal Development Without Tears*. Inter-India Publications. New Delhi (1994).

Patel, M. L.: *Development Dualism of Primitive Tribes. M.D.* Publications. New Delhi (1994).

Patel, B., Barge, C.S., Kolhe, R and Sadhwani, R.: Listening to Women Talk About their Reproductive Health Problems in the Urban Slums and Rural Areas of Baroda. In: *Listening to Women Talk about their Health Issues and Evidence from India*. Gittlesohn et al. (Ed.). Har-Anand Publications. New Delhi (1994).

Patel, S.: *Ecology, Ethnology and Nutrition: A study of Kondh Tribals and Tibetan Refugees*. Mittal Publications. New Delhi (1985).

Patel, S.: Illness, Beliefs and Health Seeking Behaviour of the Bhil Women of Panch Mahal District, Gujarat State. In: *Listening to Women Talk about their Health Issues and Evidence from India*. Gittlesohn et al. (Ed.). Har-Anand Publications. New Delhi (1994).

Patnaik, N.: Primitive Tribes-So Called.pp. 1-9. In: *Tribes and Tribal Life*. Vol I, A.K.Singh (Ed.). Sarup and Sons. New Delhi (1993).

Patrika, Newspaper Dated 26th September, 1999, (1999).

Paul, B.D.: Interview Techniques and Field Relationships. In: *Anthropology Today*. A.L.Kroeber (Ed.). University of Chicago Press. Chicago (1953).

Paul, B.D. (Ed.): *Health, Culture and Community*. Russel, Sage Foundation. New Delhi (1985).

Pelto, P.J. and Pelto, G. H.: *Anthropological Research: The Structure of* Inquiry. Cambridge University Press. Cambridge (1978).

Platt, R.S.: Environmentalism versus Geography. American Journal of Sociology, 33: 441-487 (1985).

Polgar, S.: Health and Human Behaviour: Areas of Interest Common to the Social and Medical Science. *Current Anthropology*, 3: 125-205 (1962).

Population and Demographic Profile of Rajasthan in National Scenario, Directorate of Medical and Health, Family Welfare, Rajasthan, Jaipur (1995).

Powdermader, H.: *Stranger and Friend: The Way of an Anthropologist*. Norton. New York (1966).

Prabhakar, M.: *Cultural Heritage of Rajasthan*. Panchsheel Prakashan. Jaipur (1972).

Prasad, R.R. and Kapoor, A.K.: Obstetrical Practices Among the Bhotia Tribals. *Vanyajati*, 4: 7-8 (1982).

Prasad, R.R. and Kapoor, A.K.: Illness and Its Congnitive Aspect Among the Jad Tribals of Central Himalayas. *Vanyajati*, 2: 1-4 (1982).

Prema, L and Thomas, F.: Nurtition and Health Problems Faced by Kannikar Women. In: *Dimensions of Scheduled Tribes Development in India*. P.D.Tiwari and R. S. Tripathi (Ed.). Uppal Publishing House. New Delhi (1992).

Quinn, J.A.: *Human Ecology*. Prentice Hall Book Company. New York (1950).

Quraishy, Z.B.: Gender Bias in Health and Health related aspects among slum dwellers in a Madras City. *South Asian Anthropologist*,19: 25-30 (1998).

Rajasthan-Ke- Saharia. M.L.V. Tribal Research Institute, Ghatta-Ghatti Project, Goverdhanpura, Udaipur (1965).

Rajasthan Mein Sagri Vyavastha aur Ummulan -Ek – Sarvekshan. M.L.V.Tribal Research Institute, Udaipur (1975).

Rao, D.H., Mathur, Y.N., Radhaiah.,G.,Pralhad,N.: Health and Nutritional Status of Tribals in Madhya Pradesh. *NIN.(ICMR) Report*, Hyderabad (1986).

Rene, D.: *Mirage of Health*. Harper and Row Publishers. New York (1959).

Rene, D.: *WHO Chronicle*. 23: 499 (1969).

Report of a Symposium. Alternate Approaches to Health Care. ICMR. (1976).

Report of Saharia Development Plan (Revised) 1980-85. Tribal Area Development in Rajasthan, Udaipur (1980-85).

Report of a Study Group of Health for All – An Alternate Strategy. The Working Group on Development of Scheduled Tribes During Seventh Five-Year Plan. Government of India, New Delhi (1981).

Report of National Commission on Forests and Tribals. Government of India, Ministry of Home Affairs, Department of Tribal Welfare, New Delhi (1982).

Report of the Working Group on Development and Welfare of Scheduled Tribes During Seventh Five-Year Plan. Government of India, Ministry of Welfare, New Delhi (1984).

Report of the Working Group on Development and Welfare of Scheduled Tribes During Eighth Five-Year Plan. Government of India, Ministry of Welfare, New Delhi (1987).

Richards, A.I.: *Hunger and Work in a Savage Tribe: A Functional Study of Nutrition among the Southern Bantu*. George Routeledge and Sons. London (1932).

Rizvi, S.N.H.: Health Practices of the Jaunsaris: A Socio-Cultural Analysis.pp.223-246 In: *Tribal Health: Socio-Cultural Dimensions*. Inter-India Publications. New Delhi (1986).

Robinson, S., Dill, D.B., Wilson, J.W.,and Nielsen,M.: Adaptations of White Men and Negroes to Prolonged Work in Humid Heat. *American Journal of Tropical Medicine*. 21: 261-287 (1941).

Roney, J.G., Jr.: Medical Anthropology: A Synthetic Discipline. *New Physician* 8(3): 32-33, (1959).

Roney, J.G., Jr.: Medical Anthropology: An Introduction. *Journal of the National Medical Association*, 55(2): 95-99 (1963).

Rosentock, I.M.: What Research in Motivation Suggest for Public Health. *American Journal of Public Health* (1960).

Rowney, H.B.: *The Wild Tribes of India*. B.R. Publishing Corporation. Delhi (1882).

Roy Burman, B.K.: Morbidity and Nutritional Status of the Scheduled Tribes of India. pp. 85-98 In: *Tribal Health: Socio-Cultural Dimensions*. B. Chaudhuri (Ed.). Inter-India Publications. New Delhi (1986).

Roy Burman, B.K.: *Compilation and Report. A Preliminary Appraisal of the Scheduled Tribes of India,* Office of Registrar General, New Delhi (1988).

Roy Burman, B.K.: *Encylopaedia on STs, SCs and Disadvantaged People-VolI,SeriesI*, Mittal Publications. New Delhi (1994).

Roy, S.C.: *The Birhor: A Little Known Tribe of Chottanagpur*. Man in India Office. Ranchi (1925).

Roy, S.C.: *Oraon Religion and Customs*. Man in India Office. Ranchi (1928).

Roy, S.C.: *The Hill-Bhuiyas of Orissa*. Man in India Office. Ranchi (1935).

Roy, S.C.: Magic and Witchcraft on the Chota-Nagpur Plateau: A Study in the Philosophy of Primitive Life. *Man in India*, 64 (1984).

Russel, R.V. and Hiralal.: *Tribes and Castes of Central Provinces of India*. MacMillan. London (1916).

Sachchidananda.: Socio-Cultural dimensions of tribal Health. pp. 57-69 In: *Tribal Health in India*. S.Basu (Ed.). Manak Publications. New Delhi (1994).

Sachchidananda and Prasad, R.: *Encyclopaedic Profile of Indian Tribes*. Discovery Publishing House. New Delhi (1996).

Sachchidananda .: Study of Primitive Tribes: Certain Conceptual Issues. In: *The Profiles of the Marginal and Pre Farming Tribes of Central Eastern India*, Government of West Bengal, Calcutta (1981).

Sahu, S.K.: Social Dimensions of Health of tribals in India- A Case Study of Oraons of Orissa. In: *Tribal Health: Socio-Cultural Dimensions*. B.Chaudhuri (Ed.). Inter India Publications. New Delhi (1986).

Sarkar,S.S.: *The Aboriginal Races of India*. Scientific. Calcutta (1937).

Scotch, N.: Medical Anthropology. pp.30-38. Biennial Review of Anthropolgy. B.J. Siegal (Ed.) Stanford University Press. Stanford (1963).

Seal, S.C.: *Presidential Address.* 50th Indian Science Congress. Delhi (1963).

Selltiz, C.: *Research Methods in Social Relations*, Rienhart and Winston. New York (1961).

Sen, A.K. Settlement Patterns in Rajasthan, *Man in India*,43: 215-224 (1963).

Sharma, B.D.: *Tribal Development: The Concept and Frame.*Prachi Prakashan. Delhi (1978).

Sharma, D.: *Rajasthan Through the Ages.* (Ed.) Vol. I, Rajasthan State Archin. Bikaner (1966).

Sharma, G.N.: *Social Life in Medieval Rajasthan.* Lakshmi Narain Agarwal. Agra (1968).

Sharma, M.L.: *Kota* Rajya *Ka Itihas.* Kotah Printing Press. Kota (1941).

Sharma, S.P and Sharma, J.B (Eds.).: *Culture of Indian Tribes: Concept and Identity* Vol I,Radha Publications. New Delhi (1998).

Sharma, V and Sharma, A .: The Status of Women, Fertility and Family Planning among Tribals of South Rajasthan. *Journal of Family Welfare*, 39: 20 (1993).

Sherring, M.A.: *The Tribes and Castes of Rajasthan.* Cosmo. Delhi (1975).

Sherring, M.A.: *Hindu Tribes and Castes.* Cosmo, Delhi (1981).

Singh, A.K.: *Tribes and Tribal Life* (Vol. I) Sarup and Sons. New Delhi (1996).

Singh, A.K, Gaya Pandey and Singh, P.K.: *Forest and Tribals in India*, Classical Publishing Company. New Delhi (1994).

Singh, B.: *Tribal Development: Prospect and Retrospect.* Ministry of Home Affairs, New Delhi (1983).

Singh, B.: Factors Influencing Health of Tribal Population Groups.pp.40-56. In: Tribal Health in India. S.Basu (Ed.). Manak Publications. New Delhi (1994).

Singh, B.: Primitive Tribal Groups. pp. 70-92 In: *Indian Tribes: Tribal Studies of India Series T 182. Antequity to Modernity in Tribal India.* S.C.Dube (Ed.). Inter-India Publications. New Delhi (1998).

Singh, J.P. and Vyas, N.N.: *Tribal Development: Past Efforts and New Challenges*. MLV Tribal Research and Training Institute, Udaipur (1986).

Singh, J.P., Vyas, N.N and Mann, R.S.: *Tribal Women and Development*. Rawat Publications. Jaipur (1988).

Singh, K.S.: *Tribal Movements in India* Manohar. New Delhi (1982).

Singh, K.S.: Tribal Women: An Anthropological Perspective In: *Tribal Women and Development*. K.S. Singh, N.N.Vyas and R.S.Mann (Ed.), Rawat Publications. Jaipur (1988).

Singh, K.S.: *People of India. Rajasthan, Part II, Vol. XXXVIII*, Anthropological Survey of India, Oxford University Press (1994).

Singh, K.S.: *People of India: National Series*, Vol III, *The Scheduled Tribes*, Anthropological Survey of India, Oxford University Press (1999).

Sinha, S.: The Problem of Land Alienation of the Tribal in and round Ranchi, *Tribal Welfare and Research Institute*, Ranchi (1968).

Sinha, S.P.: Traditional Tribal Medicine and Modern Health Care System.pp.151-163 In: *Tribal Health in India*. S. Basu (Ed.). Manak Publishers Private Limited. New Delhi (1994).

Smith, A.: Setting an Strategy for Health. *British Medical Journal*, 304: 376-378 (1992).

Some Facts About Rajasthan 1997. Directorate of Economics and Statistics, Rajasthan, Jaipur (1997).

Statistical Abstract Rajasthan 1995. Directorate of Economics and Statistics, Rajasthan, Jaipur (1997).

Srinivas, M.N.: *Social Change in Modern India*. Allied Publishers (1966).

Srivastava, S.K.: The Tharus - A Study in Culture Dynamics. Agra University Press. Agra (1958).

Steward, J.H.: *Basin-Plateau Aboriginal Sociopolitical Groups*. (Bureau of American Ethnology Bull.137) Government Printing Office,Washington (1938).

Survey of Sagri System. M.L.V. Tribal Research Institute. Udaipur (1978).

Swain, S.: Health, Disease and Health Seeking Behaviour of Tribal People of India. In: *Tribal Health of India*. S.Basu (Ed.). Manak Publications. New Delhi (1994).

Swain, S., Jena, S.C. and Singh, P.: Morbidity Status of the Kondh Tribes of Phulbani (Orissa). pp. 177-191 In: *Cultural and Environmental Dimensions of Health*. B.Chaudhuri (Ed.). Inter-India Publications. New Delhi. (1990).

Thakkar, A.V.: Some Pressing Problems of Primitive Tribes. *Eastern Anthropologist*. 3 (1949).

Thompson L.: The Relations of Men, Animals and Plants in an Island Community (Fiji). *American Anthropologist*, 253-67 (1949).

Thornthwaite,C.W.: The Relation of Geography to Human Geography. *Ecological Monographs*, 10:499-532 (1940).

Tiwari, D.N.: Primitive Tribes of Madhya Pradesh: Strategy For Development. Govt. of India, *Ministry of Home Affairs, Tribal Development Division*, New Delhi (1985).

Tiwari, S.K and Pandey, S.K.: Saharion Ka Ajaba Gajaba Sansar. *Daily Acharan dated 1st Jan 1989*, Hindi Growth Academy, Bhopal (1989).

Tod J.: *Travels in Western India*. Oriental Publishers, Delhi (1971).

Tod J.: *Annals and Antiquities of Rajasthan*. Vol. I & Vol. II, Oriental Reprint, New Delhi (1920).

Tripathi, C.B.: Denotified Communities of Rajasthan. *Vanyajati*, 11 (1963).

Union Primary Census Abstract for Scheduled Tribes. Tables for Tribal Population Statistics (1991).

Tylor, E.B.: *Primitive Culture*. J.Murry. London (1871).

Verma,S.C.: Problem of Drinking in the Primitive Tribes. *Eastern Anthropologist,* 12: 252-256 (1956).

Verma, B.S.: Rajasthan Ki Saharia Jati Ke Beech Mere Jeevan Ke 33 Varsha. *An Unpublished monograph*.

Vidich, A.J.: Participant Observation and the Collection and Interpretation of Data. *American Journal of Socoiology*, 60:354-360 (1955).

Vidyarthi, L.P.: *The Maler: A Study in Nature-Man-Spirit Complex of a Hill Tribe*. Bookland Private Limited. Calcutta (1963).

Vidyarthi, L.P.: *Nehru's Approach to Tribal Culture*. Special Tribal Orient. Ranchi (1965).

Vidyarthi, L.P.: *Applied Anthropology in India*. Kitab Mahal. Allahabad (1968).

Vidyarthi, L.P. and Rai, B.K.: *The Tribal Culture of India*. Concept Publishing Company. New Delhi (1985).

Vyas, N.N and Bhil, R.S.: Saharia: A Little Known Tribe of Rajasthan. *Tribe*. TRI, Udaipur 11: 20-32 (1980).

Vyas, N.N. and Chaudhuri, N.D.: *Tribals: Past and Present – Perspectives of Tribal Development in Rajasthan*, Tribal Research Institute (1968).

Vyas, N.N. and Mann, R.S.: (Eds.) *Indian Tribes in Transition*. Rawat Publications. Jaipur (1980).

Vyas,N.N. and Mann, R.S.: (Eds.) *Disease, Culture Contact and Tribals*. Tribal Research Institute, Udaipur (1980).

Warrier, P.K., P.N.K Nambiar and Pamakutty, C.: *Indian Medicinal Plants: a Compendium of 500 Species*. Orient Longmann. Madras (1995).

Wax, R. H.: *Doing Fieldwork: Warnings and Advice*. University of Chicago Press. Chicago (1971).

William, C.: Applied Anthropology in Medicine. In: A.L. Koreber (Ed.).*Anthropology Today*. Chicago University Press. Chicago (1953).

White, C.L. and Renner, G.T.: *Human Geography: An Ecological Study of Society*. Appleton-Century-Crofts. New York (1948).

White, L.: *The Evolution of Culture*. McGraw Hill. New York (1959).

Working Group Report: *Report of the working group on the Development and Welfare of Scheduled Tribes During Eighth Five Year Plan (1990-95)* Government of India, Ministry of Welfare, New Delhi (1989-90).

WHO (World Health Organization). Constitution. Geneva (1946).

WHO (World Health Organization). Environmental and Health Monitoring in Occupational Health. Technical Report No.535. Geneva (1975).

WHO (World Health Organization). Targets for Health for All. Copenhagen. (1985).

Worsley, P.: Non-Western Medical Systems. *Annual Review of Anthropology*,11: 315-348 (1982).

Young, P.V.: *Scientific Social Survey and Research.* Prentice Hall. New York (1949).

Zipf, G.K.: *Human Behaviour and the Principle of Least Effort: An Introduction to Human Ecology.*Cambridge Mass:Addison-Wesley Press (1949).

Zola .: Illness Behaviour of Working Class. In: *Blue Collar World: Studies of the American Worker.* A. Shostiks,William Comberg and Englewood Cliffs (Eds.). Prentice Hall, New York (1964).